Resorts and Ports

ONE WEEK LOAN

TOURISM AND CULTURAL CHANGE
Series Editors: **Professor Mike Robinson** *(Centre for Tourism and Cultural Change, Leeds Metropolitan University, Leeds, UK)* and **Dr Alison Phipps** *(University of Glasgow, Scotland, UK)*

Understanding tourism's relationships with culture(s) and vice versa, is of ever-increasing significance in a globalising world. This series will critically examine the dynamic inter-relationships between tourism and culture(s). Theoretical explorations, research-informed analyses, and detailed historical reviews from a variety of disciplinary perspectives are invited to consider such relationships.

Full details of all the books in this series and of all our other publications can be found on http://www.channelviewpublications.com, or by writing to Channel View Publications, St Nicholas House, 31–34 High Street, Bristol BS1 2AW, UK.

TOURISM AND CULTURAL CHANGE
Series Editors: Professor Mike Robinson *(Centre for Tourism and Cultural Change, Leeds Metropolitan University, Leeds, UK)* and Dr Alison Phipps *(University of Glasgow, Scotland, UK)*

Resorts and Ports

European Seaside Towns since 1700

Edited by
Peter Borsay and John K. Walton

CHANNEL VIEW PUBLICATIONS
Bristol • Buffalo • Toronto

Library of Congress Cataloging in Publication Data
A catalog record for this book is available from the Library of Congress.
Resorts and Ports: European Seaside Towns since 1700/Edited by Peter Borsay and
John K. Walton.
Tourism and Cultural Change: 29
Includes bibliographical references.
1. Seaside resorts–Europe–History. 2. Harbors–Europe–History. 3. Tourism–Europe--History.
I. Borsay, Peter. II. Walton, John K. III. Title. IV. Series.
GV191.48.E8R47 2011
914.062–dc23 2011027891

British Library Cataloguing in Publication Data
A catalogue entry for this book is available from the British Library.

ISBN-13: 978-1-84541-198-5 (hbk)
ISBN-13: 978-1-84541-197-8 (pbk)

Channel View Publications
UK: St Nicholas House, 31–34 High Street, Bristol BS1 2AW, UK.
USA: UTP, 2250 Military Road, Tonawanda, NY 14150, USA.
Canada: UTP, 5201 Dufferin Street, North York, Ontario M3H 5T8, Canada.

The policy of Multilingual Matters/Channel View Publications is to use papers that are natural,
renewable and recyclable products, made from wood grown in sustainable forests. In the
manufacturing process of our books, and to further support our policy, preference is given to
printers that have FSC and PEFC Chain of Custody certification. The FSC and/or PEFC logos
will appear on those books where full certification has been granted to the printer concerned.

Typeset by Datapage International Ltd.
Printed and bound in Great Britain by the MPG Books Group.

Contents

List of Figures

Acknowledgements

Permission to use the illustrations listed in the List of Figures has been given by courtesy of the person or institution named in parentheses. We are very grateful to these parties for granting this permission and for the considerable help that they have provided us in obtaining the images.

Contributors

Peter Borsay is Professor of History at Aberystwyth University, a member of the International Advisory Board of *Urban History*, and a committee member of the British Pre-Modern Towns Group. His books include *The English Urban Renaissance: Culture and Society in the Provincial Town, 1660–1770* (Oxford University Press, 1989); *The Image of Georgian Bath, 1700–2000: Towns, Heritage and History* (Oxford University Press, 2000); and *A History of Leisure: The British Experience Since 1500* (Palgrave, 2006). He is currently engaged in research on 'green' space in or on the edge of British towns (1660–1900) and on spas and seaside resorts in Britain and is preparing a book on *The Discovery of England*.

Allan Brodie is an architectural historian who works as a senior investigator in English Heritage. He is the co-author of *Seaside Holidays in the Past* (2005) and *England's Seaside Resorts* (2007), and he has also published books on the seaside heritage of Margate and Weymouth. He is also the co-author of *Behind Bars* (2001) and *English Prisons: An Architectural History* (2002). In 2011, he was the co-author of a book on the defences of the Isles of Scilly and the threat of climate change to these structures (*Defending Scilly*), and he published the section of the Salisbury and South Wiltshire Museum Catalogue that covers medieval stone sculpture.

Jan Hein Furnée is Lecturer in Modern History at the University of Amsterdam. His research focuses on urban leisure and consumption in the nineteenth century – from clubs and societies to theatres, concert halls, zoos, seaside resorts and shopping streets. His main publications in English include 'Bourgeois strategies of distinction: Leisure culture and the transformation of urban space: The Hague, 1850–1890', in S. Gunn and R.J. Morris, eds, *Identities in Space: Contested Terrains in the Western City Since 1800* (Ashgate, 2001, pp. 204–227) and 'In good company: Class, gender and politics in The Hague's gentlemen's clubs, 1750–1900', in R.J. Morris, G. Morton and B. de Vries, eds, *Civil Society, Associations and Urban Places. Class, Nation and Culture in Nineteenth-Century Europe* (Ashgate, 2006, pp. 117–138). He is a member of the editorial boards of the journals *Stadsgeschiedenis De Negentiende Eeuw* and *Geschiedenis Magazine*, and he is Secretary of the European Association for Urban History.

Fred Gray is Professor of Continuing Education at the University of Sussex, United Kingdom. He has a long-term interest in all things seaside and, in particular, the cultural history and architecture of coastal resorts in Britain and other Western countries. He is the author of *Designing the Seaside: Architecture, Nature and Society* (Reaktion Books, 2006, and issued in paperback in 2009). In 2009, he was the co-author with David Powell of a study of the economic recession and cultural regeneration in four southeast England coastal towns. He is currently working on a cultural history of the palm tree.

David Hussey is Subject Leader in History at the University of Wolverhampton. His interests are focused on trade, consumption, material culture and gender in the long eighteenth century. His publications include *Coastal and River Trade in Pre-Industrial England* (Exeter University Press, 2001); *Buying for the Home: Shopping for the Domestic from the Seventeenth Century to the Present* (Ashgate, 2008) (co-edited with M. Ponsonby); *Teaching Gender and Sexualities* (CSAP, Birmingham, 2011) (co-edited with M. Mirza); and *The Single Homemaker and Material Culture in the Long Eighteenth Century* (Ashgate, 2011, forthcoming).

Berit Eide Johnsen is Professor of History at the University of Agder in Kristiansand, Norway, and a member of the executive committee of the International Maritime Economic History Association (IMEHA). She is a graduate of the Universities of Oslo and Bergen and obtained her PhD in history in 1998 with the dissertation *Rederistrategi i endringstid 1875–1925 (Shipping Strategies in a Time of Change. The Shipping Industry of Southern Norway from Sail to Steam and Motor, from Wood to Iron and Steel. 1875–1925)*, published in 2001. Besides a number of articles and books on cultural history and different aspects of the shipping industry of southern Norway, she has also studied and written about the history of tourism in Sørlandet, southern Norway. She is currently engaged in a history research project concerning the village of Lillesand, southern Norway, 1800–2000, with an emphasis on shipping and tourism.

Simo Laakkonen works as a university lecturer at the School of Cultural Production and Landscape Studies, University of Turku, Finland. He has published several books on urban environmental history. His main interest is the environmental history of the Baltic Sea.

Louise Miskell is a Senior Lecturer in History at Swansea University and an associate editor of *Urban History*. Her publications include the co-edited collection *Victorian Dundee: Image and Realities* (Tuckwell Press, 1999); *Intelligent Town: An Urban History of Swansea, 1780–1855* (University of Wales Press, 2006) and *The Origins of an Industrial Region: Robert Morris* *and the First Swansea Copperworks, 1727–1730* (South Wales Record Society, 2010). Her current research interests focus on the relationship between industrial and recreational development in nineteenth-century Welsh towns and the impact of scientific meetings on the towns and cities of Victorian Britain.

Guy Saupin is Professor of Modern History at Nantes University, a committee member of the Research Centre of International and Atlantic History and a member of the Groupe intérêt scientifique français d'histoire maritime. His books include *Les villes en France à l'époque moderne* (Paris, Belin, 2002); *La France à l'époque moderne* (Paris, A. Colin, 2010, 2000, 2004); (ed.) *Villes atlantiques dans l'Europe occidentale du Moyen Âge au XXe siècle* (Rennes, PUR, 2006); (ed.) *Le commerce atlantique entre la France et l'Espagne à l'époque moderne* (Rennes, PUR, 2008).

Karina Vasilevska is a researcher of ethnology at the University of Latvia. She has also studied cultural anthropology at the London School of Economics and worked in the United States.

John K. Walton is an IKERBASQUE Research Professor in the Department of Contemporary History, University of the Basque Country UPV/ EHU, Leoia, Bilbao. He previously held chairs at Lancaster University, the University of Central Lancashire and the Institute of Northern Studies, Leeds Metropolitan University. He edits the *Journal of Tourism History* for Routledge and has published extensively on the histories of tourism and coastal resorts, particularly in Britain and Spain, as well as on the social history of sport, cooperation and regional identities (among other things). His most recent book is (with Keith Hanley) *Constructing Cultural Tourism: John Ruskin and the Tourist Gaze* (Bristol: Channel View, 2010).

Jason Wood has been an archaeologist and heritage consultant for over 30 years and has been Director of Heritage Consultancy Services since 1998. His previous posts have included Professor of Cultural Heritage at Leeds Metropolitan University, Head of Heritage at WS Atkins Consultants Limited and Assistant Director of Lancaster University Archaeological Unit. He has a special interest in the emerging fields of public history and the heritage of sport and leisure and serves on the executive committee of the British Society of Sports History and the International Advisory Board of the *Journal of Tourism History*. In recent years, a particular focus of his consultancy work has been associated with the regeneration of seaside resorts, particularly Blackpool and Margate.

Chapter 1

Introduction: The Resort–Port Relationship

PETER BORSAY and JOHN K. WALTON

The Specialisation Hypothesis

When, in 1943, John Betjeman compiled his volume on *English Cities and Small Towns* in the Collins *Britain in Pictures* series, he devoted separate sections to 'Ports' and 'Spas and Watering Places'. To reinforce the distinction, he declared,

> After the visit of George III to Weymouth in the eighteenth century, watering places sprang up on the coast, and they must not be confused with the sea ports, where the sea is chastened by harbour bars and docks. In watering places, everything is a preparation for playing on the edge of the sea and for looking at it.[1]

He was, excusably, wrong about both timing and causation,[2] but historians have tended to embrace this separation of functions, with resorts and ports generating discrete historiographies.[3] Underpinning such an approach has been the notion that industrialisation and rapid urbanisation brought about a greater specialisation of urban function, since towns were defined not so much by their position in a regionally defined hierarchy as by their economic role. As Penelope Corfield has argued of the eighteenth century,

> A new and more specialized terminology began to be adopted. Towns were now talked of in terms of their leading economic functions. As well as traditional concepts of market towns and ports, other places became identified as dockyard towns, manufacturing towns, spas, holiday resorts...[4]

In practice, seaports and coastal resorts grew side by side from the eighteenth century onwards, responding to the same sets of processes, of consumption as well as production, of the spread of rising living standards and aspirations, of the fashion cycle, of globalisation and increasing mobility. The emergence of seaside resorts formed an integral part of the industrialisation process rather than constituting a subsequent consequence of this long and complex sequence of developments. This is worth emphasis because it has not always been understood,[5] and it should also be stressed that these were not geographically isolated developments but often shared the same locations or adjacent ones. It is widely recognised that resorts were not necessarily, or even usually, built on virgin sites and that many had developed out of fishing settlements and ports, but such economic roles are usually described as 'decayed' and 'moribund'.[6] Some historians have acknowledged that it was not necessarily so easy to distinguish a resort from a port, since both functions could continue to operate in tandem,[7] and that the reality on the ground was far messier than the specialisation hypothesis would suggest. P. J. Waller, for example, has argued that 'seaside towns were not homogenous types ... they often combined holiday facilities with other pursuits, usually shipping and fishing', and that 'the history of pleasure resorts ... is more complicated than a story of property tycoons and corporations sniffing ozone and cashing in on an inevitable boom. One factor is evidently the potential for alternative business. A certain level of port traffic would not upset the holiday trade'.[8] In fact, a large number of railway connections to emergent resorts were built with the primary intention of developing freight traffic to commercial harbours, and the resorts, with their fluctuating, unreliable and inconvenient seasonal traffic, were the secondary beneficiaries of such initiatives.[9]

Contemporary guidebooks – which might be inclined to hide the presence of intrusive aspects of trade and commerce from potential visitors – could not conceal the obvious fact that conventional business activity mixed freely with the business of pleasure in some maritime settlements. Baedeker in 1894 pronounced Folkestone on the Kent coast 'a cheerful and thriving seaport and watering-place' and Dover, with its 'large outer tidal basin and two spacious docks' to accommodate the continental mail packets, 'a favourite bathing-place and winter-resort'; on the east coast, Lowestoft was 'a fashionable sea-bathing resort' and 'important fishing-station', and Yarmouth 'the most important town and port on the E. Anglian coast ... is also a very popular watering-place and in the summer is flooded almost daily with excursionists'.[10] An annual guide to *Seaside Watering Places* for the season 1900–1901 was typical in including among its entries the major fishing or shipping ports of Southampton, Swansea, Brixham, Falmouth, Plymouth and Grimsby, as well as many minor ones.[11] Clearly, resorts and ports were not mutually exclusive categories of settlement. This is not to say that in any particular location these functions were equally balanced or that, over time, one did not come to dominate and maybe drive out the other. Many small ports and fishing villages gradually completed a long-term transition, evolving into resorts during the eighteenth and nineteenth centuries but without necessarily losing older activities altogether. The two roles could co-exist over a considerable period of time so that any simple story of one automatically displacing the other, as the forces of specialisation kicked in, is difficult to sustain. Instead, a more nuanced and complicated account needs to be developed of how resort and port interacted with each other. The studies in this volume explore this relationship and the various transitions under way through a set of detailed case studies ranging across Britain and Europe. The term 'port' is defined broadly, embracing commercial, military, manufacturing and transport activity associated with maritime business, whereas coastal 'resorts' include a range of health and leisure functions, including pleasure boating, as well as those directly related to bathing beaches. There remains an extensive middle ground of retailing, services and infrastructure provision and maintenance which might serve port, resort and (for example) retirement functions to varying and changing degrees, with definitions being complicated by the prevalence of multiple occupations, casual, part-time and 'informal' employment, and seasonal migration in both resort and port economies.[12]

The following introduction sets out an agenda for investigating the resort–port relationship. It begins by reviewing the current historiography before examining the extent to which resorts continued to function as working ports and industrial and service centres and how far this relationship was one of conflict or co-existence. There then follows an exploration of the critical role of imagery in shaping visitor perceptions of the compatibility, or otherwise, of resort and port. How this marriage has fared over the long term is then briefly reviewed before a conclusion introduces the detailed studies contained in this volume and outlines some of the common themes that emerge among them.

Historiography

Historical surveys of ports and resorts across countries or continents have not been common, and the overall balance of studies tilts strongly towards the ports. This is not surprising since coverage at this level of generalisation has understandably highlighted the major port cities for which resort functions, where they existed, remained peripheral in every sense. A recent collection of papers on 'Western European port-cities' between 1650 and 1939, for example, hardly mentions coastal resort activity, apart from a few brief and isolated comments on sea-bathing at Portsmouth and Southsea, while O'Flanagan's ambitious survey of the port cities of Atlantic Iberia over four centuries similarly neglects coastal resort developments in (for example) Santander and San Sebastián, which come towards the end of his period, in 'second-tier' seaports, and after the themes that most interest him begin to wane from the late eighteenth century.[13] Studies of Liverpool as a city and seaport economy tend to ignore its own satellite resorts and 'marine suburbs', whether it was New Brighton on the opposite bank of the Mersey, Parkgate and its neighbours on the Wirral Peninsula or Crosby, Waterloo, and Southport to the north of Bootle's dockland.[14] The kinds of smaller ports in which resort functions became more prominent have tended to be studied at the regional or local level, as case studies or for their perceived intrinsic interest; and even then, as in recent studies 'in the round' and over long periods of the neighbouring Yorkshire resorts of Scarborough and Bridlington, the resort aspects of the urban economy have tended to be subordinated (understandably at times) to more conventional themes.[15] Substantial towns where coastal resort functions have been grafted on alongside commercial port, fishing and other economic activities, including Ostend, Boulogne, Dieppe, San Sebastián, Santander, Málaga and various Baltic ports as well as the British examples cited above, have attracted a scattered but cumulatively impressive set of historical analyses without the 'port and resort' theme developing into anything resembling a full-scale historiography.[16] There were other patterns as well. At Rimini, different kinds of industrial activity, in symbiosis with coastal tourism, proved more important to regional development than maritime activities, whereas at Bilbao the sheer intensity of commercial and industrial development displaced seaside tourism far down the Nervión estuary, beyond the city boundary, to Getxo and beyond.[17] We do, however, have a few introductory surveys of patterns of coastal resort development in its own right across broad areas of Europe, although their coverage is tilted towards the western half of the continent, and the absence of great cities among the resorts makes them less visible in national and international urban hierarchies.[18] But the working out of relationships between port and resort has mainly been a matter for local study, particularly in smaller settings. It is particularly important to escape from a prevailing set of assumptions that regards seaside tourism as simply an escape route from the decline of fishing and maritime trade as competition from the bigger ports intensified and the railways made their impact; as the case studies in this book demonstrate, the realities were much more complicated.

Co-existence or Competition?

Coastal resorts were, then, often to be found in symbiosis with fishing and commercial ports, even with associated manufacturing and import processing industries, and each function could benefit from the presence of the other, although the more exclusive resort interests were sometimes reluctant to recognise this. Naval and international passenger ports might provide distinctive assets too: the Royal Navy offering a military spectacle to holidaymakers at Southsea or on the Isle of Wight, or majestic transatlantic passenger liners passing New Brighton on their way to and from the Liverpool docks.[19] In other cases, the functions were also at arm's length: Grimsby's deep-sea fishing and commercial docks were a short train or tram ride from the popular resort of Cleethorpes in Victorian times and thereafter, and interested parties had to make a special effort to visit them, although they were certainly recognised as an additional attraction.[20] At Hastings, on the other hand, the fishing boats were drawn up on the eastern beach in front of the Old Town, and the business of fishing, with the distinctive architecture of the Net Stores, was an attraction for holidaymakers. This did not prevent twentieth-century local government from pushing it steadily further away from the 'polite' and controlled part of the resort and seeking to demolish large areas of the quaint and attractive fishing quarter, giving rise to a conflict which was reprised on many parts of the British coastline in the mid-twentieth century, with varying outcomes.[21]

The advent of holidaymakers also provided additional options for local fishermen, who could not only supply locally caught fish to new local markets but also make money out of fishing trips and what Sidmouth fishermen in the early twentieth century, as described by Stephen Reynolds, called 'frights': pleasure boating trips for individuals or parties.[22] This offered new flexibility to complex but fragile family economies during the summer. It also brought holidaymakers into direct contact with these objects of the 'tourist gaze', while helping to generate a series of jocular allusions in cartoon and story, in the humorous magazine *Punch* and elsewhere, to the resultant possibilities for hilarious mutual misunderstanding.[23]

Maritime trading and fishing were the most obvious roles that co-existed or competed with pleasure and health in the resort economy. But they were by no means the only ones. Coastal and estuarine locations, particularly ports, were magnets to industry.[24] The logistics and expense of moving bulky raw materials and finished products over land – particularly before the introduction of the railways and motorised vehicles – made access to water-based transport highly advantageous. Manufacturing industries utilising coal, metal ores, clay and sand – such as iron and copper smelting, earthenware production or glass making – were frequently located in or near ports. Such was also the case for coal-based industries focused on processing human consumables such as salt, tobacco and sugar. Exposed cliff faces made coastlines attractive sites for quarrying and mining, while the surge in demand in the late eighteenth century for lime for agricultural and building purposes meant that the margins of sea and estuary were often lined with kilns – some on a small scale, others generating impressive industrial plants where limestone could be burnt with coal or culm to produce quicklime.[25] Add to this shipbuilding and refitting, milling and the processing of fish products, and it was rarely possible for the world of the resort to be entirely isolated from that of industry.

Along some coastlines, where industrialisation was proceeding dramatically, such as in parts of north-eastern England and southern Wales, the proximity of the two worlds was hard to ignore even in the guidebooks, which generally blotted out any reference to factors which might detract from the appeal of a resort. At the beginning of the twentieth century, a coastal resort guide reported that Swansea

> has long become the chief seat of the copper smelting trade in Great Britain ... In addition to the great copper works there are extensive silver, steel, iron, tin, zinc, alkali and patent fuel works; breweries, tanneries, flour-mills, rope, timber, and ship-repairing yards. ... The climate is mostly healthy and agreeable ... but it is more or less spoilt by the smoke and noxious effluvia which, in some states of the wind, are brought down from the copper and chemical works of the neighbourhood. Nevertheless, the annual death-rate is comparatively low. There are magnificent sands and every advantage for sea-bathing.

Further west, at Burry Port, 'comfortable lodgings may be found ... within a few minutes walks of the sands ... there are copper and white-lead works, and a fine harbour, where a good shipping trade is done, the chief export being coal. ... The bathing is delightful, but there is no accommodation'.[26] The case of Swansea will be examined in detail later in this volume, but extensive as the impact of industrialisation was on the town, the combination of resort and industry was not unique. At the end of the eighteenth century, the small port of Instow, located at the confluence of the rivers Taw and Torridge, emerged as a modest sea-bathing resort for the middle class of north Devon. Within the settlement itself, any industrial activity probably retreated (though 'a massive fortress-like limekiln' survives), but immediately opposite, across the Torridge, lay the ship-building yards of Appledore – one of a number of such premises stretching along the Taw towards Bideford – while the estuary as a whole was pitted with limekilns, and potteries operated at Barnstaple, Bideford and Fremington, using clay mined at the last of these.[27]

Specialised industrial processes were supplemented at all resorts by a range of more general manufacturing and craft businesses – the building trades, food processing, brewing, blacksmiths, furniture making and the like – without which it was impossible for any town to operate. In the 1820s, it has been calculated that 33% of Brighton's workforce was employed in producing food, clothing and other manufactures, and another 11% was employed in the building trades.[28] A further 42% was engaged in providing non-accommodation services. Many of these would of course provide for the needs of visitors as well as residents. Some resorts were also regional centres, catering not only to holiday-makers but also to shoppers, entertainment seekers and the like from surrounding villages and nearby urban networks, as with Blackpool and the Fylde of Lancashire. As such, regionally orientated consumer and professional services – including educational facilities – could represent an important strand in their economies. In 1833, for example, Brighton was said to have about 90 schools for 'young ladies and gentlemen', some of whom would come from more distant parts, as well as over 50 teachers of particular skills such as music, languages and writing.[29] At Aberystwyth, the worlds of leisure and education became inextricably intertwined when the entrepreneur who was building the spectacular sea-front Castle Hotel, Thomas Savin, went bankrupt in 1866. The University College of Wales, seeking a site on which to establish itself, grasped the opportunity provided by a knock-down price and an impressive location and took over the building project. Work continued on the site during the rest of the century, the university absorbed further property along the promenade to provide student halls, and the Alexandra Halls (for women) were constructed in 1896–1898 at the northern end of the promenade; meanwhile, the Cambrian Hotel opposite the pier failed and was converted into a Calvinistic Methodist theological college from 1906. The fusion of higher education and resort location was to prove prophetic of a trend that was to bear considerable fruit in the later twentieth and early twenty-first centuries, particularly in Britain, as seaside towns sought pathways to regeneration.[30]

By the turn of the millennium, indeed, many ports and resorts, particularly in northern Europe, had been encountering difficult times for a generation. This applied particularly to smaller, more 'traditional' places, which saw commercial maritime traffic concentrating in a handful of huge container ports for distribution by road and main-line rail; although local ship-building and repairing also lost markets, inshore and middle-water fishing suffered from ever-tightening quotas and regulation as well as from intensive trawling by international fleets serving factory ships, and seaside tourism lost out to cheap and sunny competition on, particularly, Mediterranean coastlines. Not that the latter were immune to problems of their own, since resorts built as part of the post-war tourist boom in Spain or Italy were beginning to look dated (and lacking in character) and were having difficulty meeting the challenge of new destinations in the eastern or southern Mediterranean or indeed far beyond Europe. Certain kinds of leisure boating activities provided a boost to maritime tourism, particularly yachting, and the development of 'marinas' was a widespread element in coastal regeneration from the 1970s onwards, although disruptive or insensitive proposals could provoke outraged opposition in places and at periods as far apart (within England) as Brighton and Whitby in the 1970s and the early twenty-first century.[31] On a less ambitious scale, boating, yachting and recreational fishing became gradually and readily incorporated into harbour environments, more so at times than the more conventional inshore and middle-water fishing industry, which was capable of provoking complaints about smells and obstruction even as it attracted sightseers and photographers, and its installations and artefacts provided unobtrusive enhancements to the distinctiveness of seaside settings.[32] Fishing was capable of modernising itself, unobtrusively, updating its technologies and seeking re-equipment grants without provoking opposition to what might amount, cumulatively, to a transformation of the appearance and activities of a resort harbour with or without the addition of a marina.[33] All changes in port economies, whether associated with change or decline in established commercial activities or with the penetration of tourism and its often contradictory expectations about tradition and modernity into new areas, were likely to promote economic conflict, not least over what constituted appropriate levels of local government support or subsidy (through, for example, differential rates for mooring and use of services) for different economic activities; and where this type of conflict was an issue, subsidies for aspects of the holiday industry, particularly entertainment programmes, might also come into question.[34] It was not to be expected that resort and older maritime functions would dovetail and adjust simply and easily, and it was seldom the case; however, they combined in changing measure to contribute to the construction and dissemination of images of coastal tourism.

Image

Resorts depended hugely upon their image. They were a commercial product or brand and would have to convince potential consumers of the meaning and value of what they sold. From a practical perspective, many of those customers would have no experience or very limited personal experience of their chosen destination, and may even have only a restricted time – for the day-tripper scarcely any time – to acquaint themselves with their temporary habitat. Holidaymakers were highly reliant upon others to supply mental maps and frames to interpret the resorts they visited. The emphasis upon imagery was reinforced by the centrality of looking and visualisation to the seaside experience. In part, this was because resorts developed rich audience-orientated entertainments such as plays, shows and street and beach theatre. But it also owed much to the fact that visitors were part of the social drama, engaging in intensive observation of each other – what has been called auto-voyeurism – in a context that encouraged high levels of personal and group performance and display. Alongside these forms of social visualisation went acute engagement with the natural and built environment through consumption of 'views': of sea, beach, sky, cliffs, rocks and greenery; of ships, boats, yachts, piers, pavilions, terraces, hotels, public buildings and picture palaces; and of combinations of these.

The processes of branding and mental orientation on the one hand and social and environmental visualisation on the other depended upon and gave rise to a remarkably rich culture of written and visual representation, which included guidebooks, newspapers, fictional literature, drawings, cartoons, paintings, posters, photographs, postcards and, from the early twentieth century, film. Resorts also had their own soundscapes, artificial as well as 'natural' (breakers, eddies, the wind in the rigging, the cries of seabirds), ranging from the insistent and often irritating rhythms of the 'German band' or the conventional offerings of minstrels and *pierrots* to the artistically ambitious orchestras of Folkestone and Bournemouth, Scarborough or even Blackpool.[35] For the subject of this volume – the relationship between resort and port – the importance of imagery to selling and interpreting a resort poses a set of interconnected problems. There was a strong tendency for representations to focus on the needs of visitors rather than inhabitants and to emphasise the positive, recreational, healthy features of a resort and its positive, desirable qualities of 'otherness'. At the very heart of the seaside was its role as an antidote to work, to the unhealthiness of the industrial city, and to the boredom and banalities of everyday routine (although there were plenty of jokes about the alternative banality of seaside routines, particularly in the smaller resorts). There would seem little impetus, therefore, for resort imagery to portray commercial and industrial life or to reveal the working infrastructure that supported the public facade of pleasure, except where jokes could be made about the availability and quality of accommodation and other services. Advertising posters, such as those for the railways, inevitably focus their gaze on the sea, the beach, promenades, healthy bodies, recreations and fine buildings and views.[36] H.G. Gawthorn's mid-1920s London and North Eastern railway poster for Great Yarmouth and Gorleston looks across a golden beach from the Wellington pier (1853–1854, with its 1903 pavilion and reconstructed winter gardens purchased from Torquay framing the image) to the Britannia pier (1858, rebuilt in 1902) in the distance. Elegantly dressed but relaxed holidaymakers populate the picture. There is not a hint of commercial or industrial Yarmouth to the rear of the beach zone, although it was still, in the early twentieth century, the home of a great commercial port and seasonal herring fishery. In 1921, 1149 vessels fished from the town, and four years later, the Scottish curers (who landed the majority of the catch) deployed 757 boats and employed 4000 fisher girls whose skills, badinage and personal attractiveness contributed to the holiday attractions of late summer and early autumn.[37]

However, the picture was not quite as straightforward as it might seem. Visitors to resorts travelled on forms of transport – steamboat, railway and, later, motorised coach and car – that, despite being industrial in their operation and plant, were seen as part of the holiday experience and readily included in resort imagery. So also was a whole range of pleasure craft, even if some of these might normally serve commercial purposes. Nineteenth-century paintings of the coast not only included dramatic natural features such as wild seas and soaring cliffs but also fishing boats, fish markets and fishing folk and their families, with the latter usually in poses that suggest the stoicism, nobility and sometimes heroism of their subjects.[38] Hastings, like many other European coastal resorts, acquired a group of artists and photographers who were more interested in depicting seafaring, fishing and port activities than they were in portraying beaches and holidaymakers; so here, too, port and related activities were turned to account for holiday-related purposes. As Nina Lübbren shows, this was a common pattern across western Europe from the late nineteenth century, often on a much larger scale than at Hastings, with artists' attachment to traditional buildings and customs often leading them to oppose innovation. She comments that artistic visitors to Concarneau drew the line at the smell from the Breton village's sardine cannery, but such a venture might itself be thought to be an alien intrusion on artisan traditions.[39] An extreme example of an enduring mutual relationship between artists and the local maritime economy arose at St Ives in Cornwall, where a particularly assertive and influential artists' colony in the inter-war years not only successfully resisted redevelopment proposals but also recruited some local people to its ranks.[40]

Photographs and postcards of resorts, though predominantly of visitors and their pastimes, also reveal a fascination with fishing culture. Fine early photographs and/or postcards were produced, for example, of the fishermen's houses in Folkestone and Falmouth and of the remarkable scenes unfolding as a huge herring catch was unloaded, graded and gutted at Yarmouth. Such visual imagery, supported by references in the guidebooks, demonstrates that aspects of the fishing industry could be accommodated within acceptable representations of resorts.[41] This reflected the fact that by the late nineteenth century, if not before, townspeople's image of the seaside, like that of the countryside, included healthy working people engaged in traditional occupations. The industrial fishing of the long-distance trawler and drifter fleets of Grimsby, Great Yarmouth or Fleetwood might also provide educative entertainment, but visiting their docks was a separate excursion rather than an integral part of the holiday experience, as was the case in smaller, inshore and middle-water ports such as Scarborough or Whitby. At the same time as playing to the Victorian cult of work (ambiguously, because fishermen on shore were seldom models of focused, sustained physical activity), the traditional fishing industry as a tourist attraction (even as it evolved to embrace new technologies and motive power and incorporated pleasure outings and fishing trips for holidaymakers) celebrated notions of the authentic, primitive and pre-industrial and permitted middle-class visitors a degree of social voyeurism. However, fishing was seen as work quite 'other' than that experienced in the city, and those engaged in it offered an attractive spectacle to the holidaymakers, even as they landed catches for the visitors' tables as well as for distribution to other, more distant, markets. Other forms of industrial activity and employment, such as manufacture and quarrying, were not perceived to decorate the maritime scene in a similar way and rarely feature in representations of resorts. It was, therefore, by no means impossible to reconcile the image of a resort and port, but it could only be at those points where their interests coincided and did not threaten the overall sense of attractive or exciting 'otherness'.

The Long Term

This Introduction has tended to generalise across three centuries of seaside tourism development, with roots in England and the Netherlands and a steady spread across Europe since the late eighteenth century, often adapting, building on and commercialising older popular calendar customs.[42] An attempt to impose patterns of change on the understanding of complex international processes might propose that early coastal resort development set a premium on nature and comfort, seeking tranquil, healthy environments and state-of-the-art accommodation and regarding older economic activities, and existing accommodation, as necessary evils to be marginalised or remedied.[43] Such expectations persisted, alongside a necessary tolerance of crowded and makeshift accommodation at the cheaper and more popular end of the market, which generated additional income for fishing families and the local working class; but by the second half of the nineteenth century, alternative perceptions of seaports and fishing harbours as romantic repositories of national history and patriotic virtue were taking root, along with a willingness to embrace the authenticity of untidiness. We see such perceptions gathering momentum across Europe in many of the chapters which follow, alongside an eager embrace of the modernity of sunshine, freedom, the outdoors and the clean, flowing lines of the new architectures of the inter-war years, which was perhaps more a feature of specialist middle-class beach resorts such as Bexhill and Frinton in England than of places that combined the functions of port and resort, although Morecambe's Art Deco Midland Hotel overlooked a ship-breaker's yard, which was itself a tourist attraction.[44] By the time seaside regeneration initiatives were getting underway from the late twentieth century in response to a sometimes exaggerated but in many cases real and dismaying decline in the fortunes of coastal resorts, the question of how to balance the protection of heritage against the need for investment and innovation in areas where tourists and port activities came into contact was already firmly on the table and had been since several fishing quarters had been threatened with demolition in the 1930s. Conflicts on such issues are still being worked through and are discussed in several of the chapters that follow.[45]

Case Studies and Themes

This Introduction has suggested that there is a need to revisit the orthodox view of resort development. The simple story of transition from port to resort or the notion that the two represent mutually exclusive categories is difficult to sustain. A more complex and nuanced narrative is required that recognises the complexity of the resort–port relationship. The 11 essays presented in this volume seek to provide such an account. In doing so, they also raise important questions about the broader assumption that industrialisation necessarily led to urban specialisation. Most of the case studies focus upon a particular place, because though common factors emerge in how the resort–port relationship was negotiated, there are also considerable differences reflecting the unique context of each town. The birthplace of the seaside resort was England, the country which has been the subject of the most concerted research to date, and five of the studies are drawn from there. But to give geographical breadth to the analysis, two further studies are taken from Wales – a part of Britain where the seaside has received relatively little attention – and a further four from western, southern, northern and eastern Europe. The essays are arranged in a roughly chronological order, beginning with Allan Brodie's examination of the very first seaside resorts in England. The seaside was being visited for pleasure, but not bathing, in the economically advanced Netherlands from the late seventeenth century. Jan Hein Furnée's examination of the resort of Scheviningen traces its development from its origins until 1900, enjoying a far more sustained history than the Hotwells in Bristol, whose heyday as a spa, as David Hussey's study demonstrates, was over by the early nineteenth century. This was the period when the three earliest Welsh resorts, founded in the second wave of British seaside resort creations in the late eighteenth century, were going from strength to strength. The essays of Louise Miskell on Swansea and Peter Borsay on Tenby show how two originally similar resorts reacted to very different economic contexts. Brighton was a pioneer of many aspects of seaside resort development, and Fred Gray's triple-angled view of its development from the 1730s to the present day shows how complex the port–resort relationship could be over the long term. The essays by John Walton (Whitby, England), Berit Johnsen (Sørlandet, Norway), Guy Saupin

(Gijón, Spain), Simo Lakkonen and Karina Vasilevska (Jurmala, Latvia) and Jason Wood (Margate, England) all focus on the development of the resort since the late nineteenth century, when for the first time it accommodated the challenging varieties of a broader popular culture as working-class holidaymakers began to flock to some seaside locations in large numbers. This was also the period when resorts became paradoxically both a vehicle for modernism[46] and – as most of these essays demonstrate – embodiments of a deep nostalgia for the traditional world of the sea.

One of the most striking features to emerge from this collection of essays is, despite the geographical and chronological range of the resorts examined, that the resorts share common experiences. The turn to tourism as a response to the decline of a traditional economic function, such as fishing or manufacture, is clearly something shared by many resorts. This might apply in the past as well as the present, to eighteenth-century Brighton and Tenby as to early twenty-first century Gijón. But what is also evident, and has been underappreciated, is the extent to which it was possible for the recreational/health and commercial maritime functions to co-exist. In the eighteenth century at Scarborough, Margate and Bristol Hotwells, spa/resort and port operated alongside each other, and in the last of these cases, decline affected both simultaneously, suggesting a substantial degree of inter-dependence. Conflict or tension between port and resort interests – as at Scheveningen, Sørlandet, Swansea and Whitby – is a frequent characteristic of the towns studied. But the very existence of tensions suggests the continuing presence of *both* activities. Stresses and strains could be, to some extent, absorbed by spatial segregation: at Swansea, the resort moved westwards towards the 'slip' and Mumbles as the docks developed; in Whitby, the port occupied the Esk estuary while the Victorian resort grew up on top of the western cliffs; in the case of Jurmala, industrial and port functions were safely contained within neighbouring Riga; and at Brighton, the fishing business was largely relocated at nearby Newhaven and Shoreham. However, the reinvention of a fishing quarter on the promenade at Brighton in the late twentieth century or the growth of a fishing tradition in Sørlandet during the same period suggests that incorporation may be as frequent a response as segregation. From the late nineteenth century and in some cases earlier – as Scheveningen, Tenby and Whitby demonstrate – the presence of traditional fishermen and women, their families, vessels and houses became part of the visitor experience, an antidote to modernisation and industrialisation. In many respects, this depended as much upon image – a quality resorts traded heavily in – as actuality. Guidebooks and illustrative material had tended to airbrush inappropriate elements from representations of resorts, and certain recent approaches to regeneration through one-size-fits-all modernisation have sought to sweep them away. But by the late Victorian period, art, painting and photography were turning an appreciative eye to fishing folk, even if industry and the less aesthetically attractive elements of the shipping business remained taboo. In a number of resorts, such as Scheveningen and Jurmala, image also became closely linked to issues of national identity, and in the case of Britain, where the sea has played an iconic role in forming a collective consciousness, the seaside holiday may have stronger patriotic overtones than is often understood to be the case. What is clear from almost all the locations studied in this volume, in Britain and Europe, is that the relationship between resort and port is now a core element in the broader heritage industry that has become a central feature of tourism in recent decades. Resorts have been able to play the regeneration game in part by drawing upon their own leisure heritage, as the case of Margate demonstrates, but also, and to a greater extent, by imaginatively reintegrating the port into their material and cultural fabric.

Notes and References

1. John Betjeman, *English cities and small towns* (London, 1943), p. 34.
2. He was writing after E.W. Gilbert's pioneering historical geography of the British seaside ('The growth of inland and seaside health resorts in England', *Scottish Geographical Magazine*, 55, 1939, pp. 16–35) but before J.A.R. Pimlott's remarkably thorough, imaginative, and prescient *The Englishman's Holiday: a Social History* (London, 1947).
3. See for example Penelope Corfield, *The Impact of English Towns 1700-1800* (Oxford, 1982); Peter Clark, ed., *The Cambridge urban history of Britain*, Volume II, *1540–1840* (Cambridge, 2000); Martin Daunton, ed., *The Cambridge urban history of Britain*, Volume III, *1840–1950* (Cambridge, 2000).
4. Corfield, *Impact of English towns*, p. 16.
5. M. Boyer, *Histoire Générale du Tourisme, du XVIe au XXIe siècle* (Paris, 2005) ; A. Corbin, *The Lure of the Sea* (Cambridge, 1994).
6. Arthur E. Smailes, *The geography of towns* (London, 1953), pp. 50, 128.
7. J.K. Walton, *The English seaside resort: a social history 1750–1914* (Leicester, 1983), p. 47.
8. P.J. Waller, *Town, city and nation: England 1850–1950* (Oxford, 1983), pp. 135, 139; J.K. Walton, 'The seaside resorts of England and Wales, 1900–1950: growth, diffusion and the emergence of new forms of coastal tourism', in A. Williams and G. Shaw, eds., *The rise and fall of British coastal resorts* (London, 1997), pp. 21–48.
9. J.K. Walton, 'Railways and resort development in Victorian England: the case of Silloth', *Northern History*, 15 (1979), pp. 191–209.
10. K. Baedeker, *Great Britain: handbook for travellers* (Leipzig, 1894), pp. 14–15, 448–9.
11. *Seaside watering places* (London, 1900).
12. J.K. Walton, 'Seaside resorts and maritime history', *International Journal of Maritime History*, 9 (1997), pp. 125–47.

13. R. Lawton and R. Lee, eds., *Population and Society in Western European Port Cities, c. 1650–1939* (Liverpool, 2002); P. O'Flanagan, *Port Cities of Atlantic Iberia, c. 1500–1900* (Aldershot, 2008).
14. M. Hope, *Castles in the Sand* (Ormskirk, 1982); J. Liddle, 'Estate management and land reform politics: the Hesketh and Scarisbrick families and the making of Southport, 1842–1914', in D. Cannadine, ed., *Patricians, Power and Politics in Nineteenth-Century Towns* (Leicester, 1982).
15. J. Binns, *The History of Scarborough: from the Earliest Times to 2000* (Pickering, 2001); D. Neave, *Port, Resort and Market Town: a History of Bridlington* (Howden, 2000).
16. J.K. Walton, 'Tourism and politics in elite beach resorts: San Sebastián and Ostend, 1830-1939', in L. Tissot, ed., *Construction of a Tourism Industry in the 19th and 20th century: international perspectives* (Neuchatel, Switzerland, 2003), pp. 287–301; S. Pakenham, *Sixty Miles from England: the English at Dieppe, 1814–1914* (London, 1967); Y. Perret-Gentil, A. Lottin y and J.-P. Poussou (eds.), *Les villes balnéaires d'Europe Occidentale du XVIIIe siècle a nos jours* (Paris, 2008); J. Smith and J.K. Walton, 'The first century of beach tourism in Spain: San Sebastián and the "playas del norte" from the 1830s to the 1930s', in J. Towner, M. Barke and M.T. Newton, eds., *Tourism in Spain: critical issues* (Wallingford,1996), pp. 35–61; C. Pellejero Martínez, 'Turismo y economía en la Málaga del siglo XX', *Revista de Historia Industrial*, 29 (2005), pp. 87–115.
17. P. Battilani and F. Fauri, 'The rise of a service-based economy and its transformation: seaside tourism and the case of Rimini', *Journal of Tourism History*,1 (2009), pp. 27–48; J.M. Beascoechea Gangoiti, 'Veraneo y urbanización en la costa cantábrica durante el siglo XIX: las playas del Abra de Bilbao', *Historia Contemporánea*, 25 (2002), pp. 181–202.
18. W. Christaller, 'Some considerations of tourism location in Europe', *Regional Science Association: Papers XX. Lund Conference 1963* (Lund, 1964), pp. 95–105; J.K. Walton, 'The seaside resorts of Western Europe, 1750-1939', in Stephen Fisher, ed., *Recreation and the sea* (Exeter Press, 1997), pp. 36–56; Walton, 'Seaside resorts of England and Wales, 1900-1950', pp. 21–48.
19. R.C. Riley, *The Growth of Southsea as a Naval Satellite and Victorian Resort* (Portsmouth, 1972); and information on New Brighton from Harry Cameron.
20. A. Dowling, *Cleethorpes: the Creation of a Seaside Resort* (Chichester, 2005).
21. S. Peak, *Fishermen of Hastings* (St Leonard's-on-Sea, 1985); J.K. Walton, 'Fishing communities 1850-1950', in D.J. Starkey, ed., *History of the fisheries of England and Wales* (Chatham, 2000), pp. 127-37; J.K. Walton, 'Fishing communities and redevelopment: the case of Whitby, 1930-1970', in David J. Starkey and Morten Hahn-Pedersen, eds., *Bridging the North Sea: conflict and cooperation* (Esbjerg, Denmark, 2005), pp. 135–62.
22. Stephen Reynolds, *A Poor Man's House* (2nd edn, London, 1909); C. Scoble, *Fisherman's Friend* (Tiverton, 2000).
23. J.K. Walton, 'Respectability takes a holiday: disreputable behaviour at the Victorian seaside', in Martin Hewitt, ed., *Unrespectable recreations* (Leeds Centre for Victorian Studies, Leeds, 2001), pp. 176–93.
24. Gordon Jackson, 'Ports 1700–1840', in Clark, *Cambridge urban history*, II, pp. 720–3; John Hannavy, *Britain's working coast in Victorian and Edwardian times* (Oxford, 2008).
25. Richard Williams, *Limekilns and limeburning* (2nd edn, Princes Risborough, 2004).
26. *Seaside watering places*, pp. 381, 386–7.

27. R. Lauder, *Bideford, Appledore, Instow and Westward Ho!* (Bideford, 196), pp.
     4–15; J. Beara, *Appledore: handmaid of the sea* (Appledore, 1990), pp. 18–19,
     31–5 ; C. Peerce, *A field guide to the archaeology of the Taw and Torridge
     estuaries* (Bideford, 2008), pp. 28–9, 36; B. Cherry and N. Pevsner, *The
     buildings of England: Devon* (2nd edn, London, 1989), pp. 127–8, 508–9; W.
     G. Hoskins, *Devon* (London, 1954), p. 143, 337; M. Brayshay, 'Landscapes
     of industry', in R. Kain, ed., *England's landscape: the South West* (London,
     2006), pp. 134, 150-1; J.F. Travis, *The rise of the Devon resorts* (Exeter, 1991).
28. S. Berry, *Georgian Brighton* (Chichester, 2005), p. 137.
29. *An historical and descriptive account of the coast of Brighton*, reprinted (London,
     1970), pp. 172–4.
30. T. Lloyd, J. Orbach and R. Scourfield, *The buildings of Wales: Carmarthenshire
     and Ceredigion* (New Haven and London, 2006), pp. 410–18, 427-8; W. J.
     Lewis, *Born on a perilous rock: Abertsystwyth past and present* (Aberystwyth,
     1980), pp. 171–6.
31. For example, Brighton Marina Ad Hoc Campaign, *The Brighton Marina
     Outrage* (Brighton, 1974).
32. J.K. Walton, 'Seaside resorts and cultural tourism in Britain', in Marguerite
     Dritsas, ed., *European tourism and culture* (Athens, 2007), pp. 195–208.
33. Peter Smith, *From the sma' lines and the creels to the seine net and the prawns*
     (Leven, 2001), for examples from the Fife coast in Scotland.
34. J.K. Walton, *The British Seaside: Holidays and Resorts in the Twentieth Century*
     (Manchester, 2000), chapter 6.
35. R. Roberts, 'The Corporation as impresario: the municipal provision of
     entertainment in Victorian Bournemouth', in J.K. Walton and J. Walvin, eds.,
     Leisure in Britain, 1780–1939 (Manchester, 1983), pp. 137–57.
36. Beverley Cole and Richard Durack, *Happy as a sand boy: early railway posters*
     (London, 1990); Tony Hillman and Beverley Cole, *South for sunshine: Southern
     Railway publicity and posters, 1923–1947* (Harrow Weald, 1999); Aldo Delicata
     and Beverley Cole, *Speed to the west: Great Western publicity and posters, 1923–
     1947* (Harrow Weald, 2000).
37. Beverly Cole and Richard Durack, *Railway posters 1923-1947* (London, 1992),
     p. 130; Colin Tooke, *Great Yarmouth and Gorleston: beside the sea* (Great
     Yarmouth, 2001), pp. 31–3; *Great Yarmouth: industrial history and sites of
     interest* (Great Yarmouth, n.d.); Charles Lewis, *The Yarmouth herring fishery*,
     Norfolk museums information sheet (1988).
38. Christiana Payne, *Where the sea meets the land: artists on the coast in nineteenth-
     century Britain* (Bristol, 2007); Charles Hemming, *British painters of the coast
     and sea: a history and gazetteer* (London, 1988).
39. N. Lúbbren, *Rural artists' colonies in Europe, 1870–1910* (Manchester, 2001),
     pp. 65, 152, 167.
40. Tom Cross, *Painting the Warmth of the Sun: St Ives Artists 1939–1975* (Tiverton,
     2008); David Tovey, *Pioneers of St Ives Art* (Tewkesbury, 2008); Michael Bird,
     The St Ives Artists (Aldershot, 2008).
41. Hannavy, *Britain's working coast*, pp. 37, 60, 122-3.
42. Corbin, *Lure of the Sea*, understates the innovatory role of England in the
     process. He is not helped by his failure to understand and make effective use
     of English language sources.
43. A. Brodie and G. Winter, *England's seaside resorts* (Swindon, 2007).
44. F. Gray, *Designing the Seaside* (London, 2006).
45. J.K. Walton and P. Browne, eds., *Coastal regeneration in English resorts – 2010*
     (Lincoln, 2010) provides an overview of current issues in Britain.

46. Lara Fiegel and Alexandra Harris, eds., *Modernism on sea: art and culture at the British seaside* (Witney, 2009); but see also J.K. Walton, 'Architecture, history, heritage, and identity: built environment, urban form and historical context at the seaside', *International Journal of Regional and Local Studies*, Series 2, 3 (2007), pp. 65–79.

Chapter 2

Towns of 'Health and Mirth': The First Seaside Resorts, 1730–1769

ALLAN BRODIE

Three centuries ago, organised sea bathing began to appear in England. For these first brave bathers, access to the sea could be enjoyed wherever there was a convenient beach and a farmhouse or inn to provide them with food and accommodation. They were seeking improved health, but if they wished to enjoy the society of other bathers, the facilities of a town were required. By the 1730s, there are a number of references to people bathing at ports as different as Scarborough, Liverpool, Brighton and Margate. Initially content to use existing facilities, increased numbers of bathers soon began to transform the life and environment of ports, and by 1769, belief in this new activity was sufficient to allow the first substantial investment in developments outside the area of the original settlement.

New behaviours, new activities and new architecture began to transform coastal towns. Between 1730 – the earliest reference to a theatre troupe visiting Margate during the summer – and 1769 – and the construction of Cecil Square in the same town – the growing number of visitors prompted a change in the tone of these towns from hard-working, religious communities to places where manners from London and the inland spas created a prevailing air of irreverent frivolity. Dr Richard Pococke, usually a keen observer of harbours and their activities, scarcely mentions the commercial life of Margate in 1754:

> This is a fishing town, and is of late much resorted to by company to drink the sea water, as well as to bathe; for the latter they have the conveniency of cover'd carriages, at the end of which there is a covering that lets down with hoops, so that people can go down a ladder into the water and are not seen ...[1]

Such sources demonstrate that after the mid-eighteenth century, the resort function began to rise. This chapter will examine the process of transforming ports into resorts from the dawn of the seaside resort to the point where it was becoming clear that the long-term future for many of these towns lay primarily in health and pleasure rather than maritime commerce.

Sea Bathing and Ports in the Early Eighteenth Century

The story of the seaside dates back to the Tudor and Stuart period. Leland's *Itineraries* (*c.* 1540) records a patient sent to a coastal village for the sake of his health; sea bathing was used for some conditions by the end of the sixteenth century; and Henry Manship refers to doctors in Cambridge in 1619 who sent patients to Great Yarmouth 'to take the air of the sea'.[2] At Scarborough, sea water was drunk as a cure to supplement the spring water being imbibed at the spa, and by 1660, Dr Robert Wittie, in his book on Scarborough Spa, noted that bathing in sea water had cured his gout.[3] By *c.*1700, Sir John Floyer, the leading advocate of cold-water bathing, was proclaiming that the sea could act as a huge cold bath to cure many ills.[4]

Scientists and medical writers were becoming enthusiastic about sea bathing, and there are indications from personal documents that people were taking to the sea by the start of the eighteenth century. In the 'Great Diurnal' of the Lancashire landowner Nicholas Blundell (1669–1737), his family sought medical treatment by visiting doctors and various spas, making a pilgrimage to Holywell in north Wales and bathing in the sea close to their home at Little Crosby, near Liverpool. Blundell's first recorded dip in the sea occurred on 5 August 1708: 'Mr Aldred & I Rode to the Sea & baithed ourselves'.[5] No medical conditions were mentioned, and therefore, he may have bathed for pleasure during hot weather. A year later, Blundell's children bathed in the sea to cure 'some out breacks'.[6] In Lincolnshire, Mrs Massingberd of Gunby described in a letter dated 2 May 1725 that 'Sr Hardolf Wastnage & his lady come in Whitsun week to a farmhouse in this neighbourhood to spend three months in order to bath in ye sea'.[7] Again, this is bathing on a suitable stretch of coastline, independent of a town; Blundell did not need facilities since he was near his home, and Sir Hardnolf Wastnage relied on an accommodating farmer.

By the early eighteenth century, some accounts of sea bathing make it clear that if a visitor wanted lodgings or entertainment, he or she would head for a coastal port. A bath-house was mentioned in a Liverpool rate assessment as early as 1708, and in 1718, Samuel Jones, a customs officer at Whitby, wrote a poem praising the water of the spa and the sea for curing jaundice:

> Here such as bathing love may surely find
> The most compleat reception of that kinde;
> And what the drinking cannot purge away
> Is cured with ease by dipping in the Sea.[8]

In August 1721, Nicholas Blundell escorted an acquaintance 'to Leverpoole & Procured him a Place to lodg at & a Conveniency for Baithing in the Sea'.[9] Chadwick's map of 1725 shows that the riverfront of

Liverpool was already dominated by docks, and there was no suitable location in the town centre for bathing. Instead, this may have taken place at the northwestern extremity of the riverfront where John Eyes' map of 1765 shows a bath-house. A detached building on this same site appears in a 1728 engraving of Liverpool produced by Samuel and Nathaniel Buck, perhaps providing the earliest depiction of a waterfront bath-house.[10]

In 1734, sea bathing was recorded in Scarborough's earliest guidebook:

> It is the custom for not only gentlemen, but the ladies also, to bathe in the seas; the gentlemen go out a little way to sea in boats (called here 'cobbles') and jump in naked directly: ... The ladies have the conveniency of gowns and guides. There are two little houses on the shore, to retire for dressing in.[11]

A spring was discovered at the port in *c.*1626, and by the 1660s, it was a well-established spa where sea bathing and drinking sea water were added to its range of treatments.

By 1730, Margate was sufficiently busy to attract a theatre company from Canterbury to perform during the summer. The earliest reference to a seaside bath-house dates from 1736 when an advertisement appeared in a Kent newspaper:

> Whereas Bathing in Sea-Water has for several Years, and by great numbers of People, been found to be of great Service in many Chronical Cases, but for want of a convenient and private Bathing Place, many of both Sexes have not cared to expose themselves to the open Air; This is to inform all Persons, that Thomas Barber, Carpenter, at Margate in the Isle of Thanett, hath lately made a very convenient Bath, into which the Sea Water runs through a Canal about 15 Foot long. You descend into the Bath from a private Room adjoining to it.[12]

It was such a success that he soon constructed a second larger bath that could be filled with sea water regardless of the state of the tide.[13] In 1736, one visitor to Brighton described how

> We are now sunning ourselves on the beach, at Brighthelmstone, and observing what a tempting figure this island made formerly in the eyes of those gentlemen who were pleased to civilize and subdue us. The place is really pleasant: I have seen nothing in its way that outdoes it. ... My morning business is bathing in the sea, and then buying fish: the evening is, riding out for air; viewing the remains of old Saxon camps; and counting the ships in the road, and the boats that are trauling.[14]

These first resorts were mostly small working towns or as John Byng snootily described them: 'fishing holes'.[15] In the early eighteenth century, John Macky saw Margate as 'a poor pitiful Place', and Lewis recorded that it was 'irregularly built, and the Houses generally old and low'.[16] The earliest ports that added a resort function had unpretentious buildings set in small plots, separated by narrow streets, and the buildings were predominantly vernacular in style and materials. Something of the atmosphere of a proto-resort can still be experienced in the old town at Margate, in the Laines at Brighton, or behind the harbour at Scarborough. Some modern coastal towns can serve as analogies for early resorts. A number of Cornish fishing towns evoke the character of early eighteenth-century Margate with its narrow streets, modest houses and small harbour. Similarly, some settlements on the Kent and Sussex coastline seem to offer parallels for the eighteenth-century seafront of Brighton or Hastings, with small fishing boats sitting on a shingle beach and a haphazard arrangement of houses along the seafront. Liverpool is, of course, an exception as a major national settlement with a large-scale port that eclipsed the modest sea-bathing function by the early nineteenth century, and Portsmouth underwent a similar process. Quebec House, a bath-house of 1754, survives near the docks, but by the nineteenth century the resort function had been exported a mile along the coast to Southsea.[17]

A number of these coastal communities including Tenby, which is discussed later in this volume, faced economic difficulties, and the new interest in sea bathing came to their rescue. Margate had prospered as the base of a fishing fleet and as an outlet for agricultural produce from the Isle of Thanet. In 1586, William Camden described the people of Thanet as 'excessively industrious, getting their living like amphibious animals both by sea and land'. Depending on the time of year, 'they make nets, catch codd, herrings and mackerel, &c. make trading voyages, manure their land, plough, sow, harrow, reap, and store their corn, expert in both professions'.[18] By the early eighteenth century, parts of the town's fishing industry had gone into decline, and in 1736, some people involved with the fishery in the North Sea had been forced to abandon their occupation altogether.[19] Brighton was once a town with fishing and cargo fleets launched from the beach, and this activity continued even after the first aristocratic sea bathers arrived. However, it is clear that fishing was in rapid decline from the 1660s and that the town's population decreased significantly during the late seventeenth and early eighteenth centuries.[20] In the 1730s, not all coastal towns were suffering economically, but for those that could attract visitors, this new activity provided a welcome addition to the local economy.

Changing Activities and Behaviours at the Earliest Resorts

The presence of visitors began to transform the behaviours and activities taking place in towns that had previously been dominated by commerce and fishing. 'Scarborough A Poem', published in 1732, celebrated a town of 'Health and Mirth' that apparently rivalled Bath.[21] Primarily because of its spa, Scarborough had been attracting visitors probably since the 1620s and therefore had the range of facilities required by its aristocratic clientele.[22] By the 1730s, it had a circulating library and an assembly room, which was described as 'a noble, spacious building, sixty-two Foot long, thirty wide, and sixteen high' with a 'Musick-Gallery' and attached card rooms.[23] A poem published in 1734 described the gaiety of the assembly room:

> When Night to Vipont's the Assembly calls,
> Engaged in Play, or lur'd to stately Balls,
> While the soft Song, and artful Dance, conspire,
> To sooth the Soul, and raise an amourous Fire.[24]

This poem appeared in a 'miscellany' of poems and prose published annually between 1732 and 1734. Much of the content suggests that a major attraction of the nascent seaside resort was the presence of beautiful, scantily dressed women. One poem focuses on the appearance of women as they emerged from the sea in thin, clinging garments:

> 'Tis here each Morn (while his full Bosom heaves)
> The green-ey'd God, the bathing Fair receives;
> With swelling Pride he presses round her Charms,
> Clasps her white Neck, and melts within her Arms.
> Now loosly dress'd the lovely Train appears,
> And for the Sea, each charming Maid prepares,
> See kindly clinging the wet garment shows,
> And evry Fold some newer Charms disclose.[25]

Other poems refer to nymphs rising from the sea and one, necessarily anonymous poet, fantasised about what it would be like to be a flower on the bosom of Miss D–rc–.[26] This suggests that something of the more licentious manners of the spas was beginning to be witnessed in hard-working, often quite religious coastal communities. Celia Fiennes described at length the presence of Quakers at Scarborough when she visited in 1697; unfortunately, there does not appear to be any contemporary evidence of their reaction to this new influx.[27]

Scarborough's earliest guidebook lists the people who stayed at the resort in 1733, including a number of dukes and earls, people with the time and wealth to reside at the seaside. Margate's ease of access from London helped to shape the type of visitors it attracted. Like Scarborough and Brighton, it was frequented by some aristocrats but not in the same numbers. Instead, guidebooks, songs and poems refer, not always flatteringly, to the social range of Margate's visitors. This became more obvious after steamers were introduced in the early nineteenth century, but there was already evidence of Margate's social diversity in the 1760s, when George Keate wrote that

> The decent tradesman slips from town for his half crown, and strolls up and down the Parade as much at ease as he treads his own shop. His wife, who perhaps never eloped so far from the metropolis before, stares with wonder at the many new objects which surround her ... The farmer's rosy-cheeked daughter crosses the island on her pillion, impatient to peep at the London females ... The Londoner views with a disdainful surprise, the awkward straw hat, and exposed ruddy countenance of the rustic nymph; who in turn scrutinizes the inexplicable coiffure of her criticiser...[28]

Scarborough was a well-developed resort by the 1730s as it had the infrastructure of a spa town, and in the second quarter of the eighteenth century, other ports began to add new facilities in imitation of those found in the more sophisticated spas. The process of development was gradual as local entrepreneurs became confident enough to make modest investments. Circulating libraries were established in any suitable building, and guidebooks to early resorts show that they were usually contained in standard houses. At early resorts, the need for an assembly room was initially recognised by enterprising innkeepers. In 1763, an assembly room was established in Margate at the New Inn on the Parade, and in 1772, Thomas Hovenden held his first assembly at the Swan Inn in Hastings in 'a suitable room, with a gallery for music'.[29] Brighton was able to sustain two rooms. Assemblies had taken place at the Castle Tavern since 1754, but a prestigious, new façade designed by John Crunden in *c.* 1776 suggests substantial growth in its business.[30] In 1761, John Hicks of the Old Ship Inn opened new assembly rooms.[31] Johanna Schopenhauer, writing later in the eighteenth century, described both:

> The Assembly rooms are in two taverns or inns, The Castle Tavern and The Old Ship Tavern. In the first, one may play cards and there is a coffee house with a billiard table and that sort of thing. The second is similar but has the advantage of accommodation for visitors although we thought the reception inferior to what we had met elsewhere in England. The rooms of both places consist, as do those in Bath, of a dance hall and several adjoining rooms for playing cards, taking tea and making conversation. All are prettily decorated and usefully furnished.[32]

Early assembly rooms had a certain architectural ambition; the same could not be said about the first theatres. Players from Canterbury came to Margate in 1730 and probably performed in an adapted building. In 1754, William Smith's company was able to use a permanent theatre, a converted barn in the Dane, and in 1771, a theatre was created in a stable at the rear of the Fountain Inn.[33] By the 1760s, Brighton had a makeshift theatre in a barn on the north-west corner of Castle Square. In 1788, Blackpool still had a similarly rudimentary facility:

> Beauty displays itself in the dance, and the place is dignified with a Theatre; if that will bear the name which, during nine months in the year, is only the threshing floor of a barn.... Rows of benches are placed one behind another, and honoured with the names of pit and gallery; the first two shillings, the other one. The house is said to hold six pounds; it was half filled.[34]

The earliest purpose-built theatres appeared at seaside resorts from the 1770s, once resorts were sufficiently large and popular to make the investment worthwhile. The first performance at the future Theatre Royal in Weymouth was in 1771, and at Brighton, a purpose-built playhouse was erected in North Street in 1773 by Samuel Paine, though it was superseded in 1789 by a larger one in Duke Street.[35] By 1778, a theatre, described as 'neat, and well adapted to its destination', had been built at Great Yarmouth.[36]

Accommodating Visitors: Transforming the Townscape of Resorts

Entertainment facilities in early resorts were initially small-scale and often rudimentary as visitor numbers were small and uncertain. With an increasing clientele and, more important, a certainty that people would return each year, entrepreneurs began to invest in more ambitious schemes. This is evident in all forms of entertainment but is particularly obvious in the accommodation being offered at resorts.

The first visitors had to find long-term accommodation in the homes of the indigenous population, with existing inns only being suitable as accommodation on their arrival. In 1736, a visitor to Brighton noted that 'as the lodgings are low, they are cheap; we have two parlours, two bed chambers, pantry, &c. for five shillings a week'.[37] At Worthing, the first holidaymakers lodged in a farmhouse while visitors to Margate in 1763 stayed in 'small but neat' houses.[38] Fanny Burney described the house she lodged in at Teignmouth in 1773: 'Mr Rishton's House is not in the Town, but on the Den, which is the mall here: it is a small, neat, thatched & white Washed Cottage neither more nor less. We are not a hundred yards from the sea, in which Mrs Rishton Bathes every morning'.[39]

Each of the earliest resorts had their own architectural character, reflecting their history, local traditions of construction and available building materials. In 1761, Anthony Relhan described Brighton as a town of six principal streets, many lanes, and some places surrounded with houses, called by the inhabitants squares. The great plenty of flint stones on the shore, and in the cornfields near the town, enabled them to build the walls of their houses with that material, when in their most impoverished state; and their present method of ornamenting the windows and doors with the admirable brick which they burn for their own use, has a very pleasing effect.[40]

At Margate, a handful of buildings survive to provide a glimpse of the town before its transformation into a resort. The Tudor House in King Street is a sixteenth-century, continuous-jettied, close-studded house, the home of a wealthy citizen, and a number of other timber-framed buildings in King Street appear to have been re-fronted in the eighteenth century. Other readily available building materials were also used. In King Street, there is a two-storied house dating from the 1680s, faced in knapped flint with brick dressings, and in Lombard Street, a brick building of the early eighteenth century is decorated with small brick arches and pilasters. Some visitors to seaside towns may have been intrigued by vernacular styles, but early tourists soon abandoned romantic notions of living in picturesque cottages in favour of more space and greater comfort in their lodgings, though there was a limited revival of interest in quaint living at Sidmouth in the early nineteenth century. The polite architectural style and non-vernacular materials were associated with the metropolis and therefore sophistication; as early as 1698, Celia Fiennes noted that Liverpool consisted of 'mostly new built houses of brick and stone after the London fashion'.[41]

Residents began to build or rebuild houses to make them more appealing to the growing number of visitors. One of Weymouth's early guidebooks suggested that 'the inhabitants by such an influx of money have been encouraged to rebuild, repair, and greatly enlarge the town, which in little more than twenty years has undergone a considerable transformation'.[42] Relhan documented the start of this process at Brighton: 'The town improves daily, as the inhabitants encouraged by the late great resort of company, seem disposed to expend the whole of what they acquire in the erecting of new buildings, or making the old ones convenient'.[43]

The early guidebook to Weymouth used the word 'transformation', but most of the new development there before the 1780s was within the area of the pre-existing settlement. The first phases of the transformation from a port to a resort at other coastal towns also involved a similar process of modification. Initially, the impact of visitors was limited to superficial changes, with existing houses being tidied up, and perhaps extended, but by the mid-eighteenth century, the fabric of streets was beginning to undergo some significant remodelling. In All Saints Street in Hastings, the original pattern of medieval timber-framed houses is clear, but some have been replaced by taller, more regular, brick Georgian houses. Since this early phase of transformation took place within the footprint of the established town, development was necessarily constrained by the pre-existing fabric of the port.

In addition, no one in early resorts could have been confident that the vogue for sea bathing would last; there was a hesitancy to invest in large-scale schemes. However, by the 1760s, a handful of people were prepared to erect more ambitious buildings for their own use. East Cliff House (1760–1762) was built on part of the East Fort at Hastings for Edward Capell, a Shakespeare scholar who spent his summers at the resort.[44] In 1766, Captain John Gould, a wealthy tea planter from India, returned to England and settled in Margate in India House, reputedly a copy of his home in Calcutta. The most prestigious of these early houses was Marlborough House on the Steine in Brighton, built between 1765 and 1769 as a large Georgian townhouse.[45]

By 1769, the pace of seaside development had quickened. Cecil Square in Margate (see Figure 2.1), laid out by 'Mr Cecil', Sir John Shaw, Sir

Figure 2.1 Cecil Square, Margate. This was the social heart of the rapidly growing resort from the late 1760s onwards and stood overlooking the Old Town, where earlier visitors had resided.

Edward Hales and several other gentlemen, marked a turning point in the story of the seaside since it was a major speculative venture based on London money. The undated 'A Summer Trip to Margate' described it: 'The new-square, which is a large one, principally built by Mr Cecil, an eminent attorney in Norfolk-street, in the Strand, consists of some very handsome houses, intended for the reception of the nobility and gentry'.[46] The new square also contained a row of shops as well as the purpose-built assembly rooms and circulating library that had been erected beside Fox's Tavern. Cecil Square symbolically breached the form of the historic area of the old town much as Queen Square marked, qualitatively, a new phase of development at Bath in the 1720s.[47] It also was the first case of businessmen collaborating to develop a large scheme and was the first new square built at a seaside resort. At other resorts, terraced houses began to be built along the seafront. In Weymouth, the first of a series of terraces was constructed to the north of the historic town in the 1780s, and by the early nineteenth century, long terraces and crescents, with houses intended to accommodate visitors during the summer, were being erected at resorts all around the coast of England.

Ports versus Resorts

> On a large Hill's Descent, stands the fair Town,
> Well built and neat, 'till now but little known.
> In Ages past, for Shipping only fam'd
> Possess'd by Mariners, and seldom nam'd,
> But now, of health and Ease the Source proclaim'd.
> Swiftly its praises round the Nation flew,
> The Nation, ever fond of something new,
> To taste its Virtues, in vast Concourse drew.[48]

John Setterington's view of Scarborough (see Figure 2.2), published in 1735, is often cited as the first illustration of sea bathing in England. This scene occupies the left half of the engraving, but the right-hand section, which depicts a busy commercial port, is usually ignored. Like the 1732 poem celebrating Scarborough 'the fair Town', Setterington recorded both the commercial activity and the new fad for 'health and Ease'. A similar blend of old and new appears in the famous view of Brighton beach where John Awsiter's bath-house of 1769 is illustrated alongside the working boats on the beach (see Figure 2.3).

To modern observers there may appear to be a conflict between port and resort functions, and the distinction would be obvious in a modern settlement through the zoning of two such different functions. However, in the eighteenth century, the concept of the resort did not exist as a separate entity, and writers saw sea bathing as an addition to the existing activities of a port. R.R. Angerstein upon visiting Britain in the

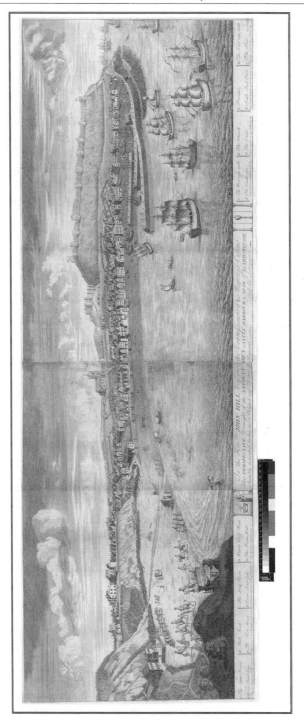

Figure 2.2 The Perspective Draught of the Ancient Town, Castle, Harbour and Spaw of Scarborough, John Setterington, 1735. This wide panorama shows a lively port with the nascent resort functioning in the foreground.

Figure 2.3 Awsiter's Baths, Brighton. Although Brighton was already the haunt of aristocrats, this view of the beach shows the basic nature of many of the early facilities.

mid-1750s described Weymouth's and Scarborough's new bathing practices alongside their commercial activity.[49] Weymouth's first bathing facilities were concentrated on the harbour-side, in the form of bathing huts and later a bath-house.[50]

The balance between port and resort has changed markedly since the eighteenth century. Some of the first resorts retain their commercial function, but at others, tourists now dominate the architecture and economy of the town. Brighton with its beach-launched fleet succumbed to the pressure of visitors. Holidaymakers came to dominate the seafront and the heart of many towns, and where they encountered existing commercial interests, there could be friction. Social tensions existed on the Steine at Brighton between 'the fishermen ... the rough sons of Peter' and 'the female anglers, who are baited with all the allurement of fashion and gaiety'.[51] Inevitably, the fishermen lost this power struggle. At Margate, the visitor was also triumphant, but at Hastings, the fleet of fishing boats and the netlofts have persisted, albeit relegated to a small area at the east end of the beach by municipal pressure and modern visitor attractions.[52]

Not all successful resorts lost their port function. Scarborough and Weymouth are still home to fishing fleets, with the latter also accommodating a large marina. At both, the harbour is a focal point for the town, but at Great Yarmouth, it is possible for tourists to enjoy their holiday while oblivious to the historic port and town on the river a few hundred metres behind the seafront. At some coastal towns, the balance between the resort and port function tipped towards commerce. Dover, a prosperous resort by the nineteenth century, is now dominated by its coastal dual carriageway, cargo docks and passenger terminals.

By the end of the eighteenth century, resorts began to be created independent of existing ports as confidence in the idea of the seaside holiday proved sufficiently robust to encourage investors to create entirely new settlements. Examples include Hothamton (now part of Bognor Regis), which opened in 1791, and Hayling Island, built in the 1820s. Both were unsuccessful, but St Leonards, founded in 1828, was a success, no doubt assisted by its proximity to Hastings. A port was no longer a pre-requisite for a seaside resort, but for the first century of the seaside holiday in England, the working coastal town was invariably the 'port of choice' for the fashionable sea bather.

Notes and References

1. Richard Pococke, *The travels through England of Dr Richard Pococke*, ed. James Joel Cartwright, (2 volumes, London, 1888-9), II, p. 86.
2. John Leland, *The Itinerary of John Leland*, ed. L Toulmin Smith, (4 vols, London, 1964), IV, pp. 43–4; H. Manship, *The History of Great Yarmouth* (Great Yarmouth, 1854), p. 104; Allan Brodie and Gary Winter, *England's Seaside Resorts* (Swindon, 2007), pp. 9–11.
3. R Wittie, *Scarbrough-spaw: or A description of the natures and virtues of the spaw at Scarbrough Yorkshire* (York, 1667), p. 172.
4. Sir John Floyer, *The Ancient ΨΛΣ' [Psykhrolysia] Revived: Or, An Essay to Prove Cold Bathing Both Safe and Useful* (London, 1702), p. 191.
5. Nicholas Blundell, *The great diurnal of Nicholas Blundell of Little Crosby, Lancashire*, ed. J. J. Bagley (3 vols., The Record Society of Lancashire and Cheshire 1968–72), I, p. 181.
6. Ibid, I, p. 225.
7. R. M. Neller, *The growth of Mablethorpe as a seaside resort 1800-1939* (Mablethorpe, 2000), p. 13, citing Lincolnshire Record Office LAO, MASS 13/16.
8. Henry Peet, *Liverpool in the reign of Queen Anne, 1705 and 1708* (Liverpool 1908), p. 56; Samuel Jones, *Whitby a poem* (York, 1718).
9. Blundell, *The great diurnal*, III, p. 52.
10. J. Chadwick, *The map of all the streets & alleys within the town of Liverpool …'* (1726); John Eyes, Plan of Liverpool docks (1765); R. Hyde, *A prospect of Britain: the town panoramas of Samuel and Nathaniel Buck* (London, 1994), plate 39.
11. Anon, *A journey from London to Scarborough* (London, 1734), p. 36.
12. *The Kentish Post, or Canterbury News Letter*, 14 July 1736, cited in J. Whyman *The early Kentish seaside* (Gloucester, 1985), p. 160.
13. *The Kentish Post, or Canterbury News Letter*, 27 April 1737, cited in Whyman *The early Kentish seaside*, p. 161.

14. J. Evans, *Recreation for the young and the old. An excursion to Brighton, with an account of the Royal Pavilion, a visit to Tunbridge Wells, and a trip to Southend* (Chiswick, 1821), p. 37.

15. C. Bruyn Andrews, ed, *The Torrington diaries 1781–1794* (4 vols., London, 1934–6), I, p. 87.

16. John Macky, *A Journey through England,* (3 vols., London, 1714–23), I, p. 50; John Lewis, *The history and antiquities . . . of the Isle of Thanet* (London, 1736), p. 123.

17. J. Webb et al., *The spirit of Portsmouth: a History* (Chichester, 1989), pp. 149–53; B. Stapleton, 'The Admiralty connection: port development and demographis change in Portsmouth, 1650–1900', in R. Lawton and R. Lee, eds., *Population and society in Western European port cities, c. 1650–1939* (Liverpool, 20002), pp. 231–2, 236.

18. William Camden, *Britannia: or, a chronological description of the flourishing kingdoms of England, Scotland, and Ireland, and the islands adjacent; from the earliest antiquity* (London, 1806), p. 316.

19. Lewis *Isle of Thanet*, p. 33.

20. S. Berry, *Georgian Brighton* (Chichester, 2005), pp 2–5, 10–11.

21. Anon, *The Scarborough miscellany* (London, 1732), pp. 1ff.

22. W. Sympson, *The history of Scarborough-Spaw* (London, 1679), pp. 5–6.

23. Anon, *A journey from London to Scarborough* (London, 1734), pp. 38–9.

24. Anon, *Scarborough miscellany*, p. 1.

25. Ibid., p. 3.

26. Ibid., p. 51.

27. Christopher Morris, ed., *The illustrated journeys of Celia Fiennes* (London, 1984), p. 101.

28. G. Keate, *Sketches from nature; taken, and coloured, in a journey to Margate* (2 vols., London, 1779), I, pp. 104–5.

29. J. A. Lyons, *Description of the Isle of Thanet and the town of Margate* (London, 1763), p. 16; G. E. Clarke, *Historic Margate* (Margate 1975), p. 76; J. Manwaring Baines, *Historic Hastings* (St Leonards on Sea, 1986), pp. 304–5; W. G. Moss, *The history and antiquities of the town and port of Hastings,* (London, 1824), p. 168.

30. Howard Colvin, *A biographical dictionary of British architects 1600–1840* (New Haven, 1995), pp. 281–2 (a plaque on the building dates the façade to 1766); J. K. Walton, *The English seaside resort: a social history, 1750–1914* (Leicester, 1983), p. 158.

31. Berry, *Georgian Brighton*, p. 27.

32. Johanna Schopenhauer, *A lady travels: journeys in England and Scotland from the diaries of Johanna Schopenhauer,* trans. and ed , Ruth Michaelis-Jena and Willy Merson, (London, 1988), p. 133.

33. Malcolm Morley, *Margate and its theatres* (London, 1966), pp. 12–17.

34. Sue Berry, 'Myth and reality in the representation of resorts' *Sussex Archaeological Collections,* 140 (2002), pp 97–112, at p. 105; W. Hutton, *A description of Blackpool in Lancashire; frequented for sea bathing* (S.I. Peneverdant Publishing, 1788, reprint 1995), pp. 37–8.

35. Yvette Staelens, *Weymouth through old photographs* (Exeter, 1989), p. 31; William Lee, *Ancient and modern history of Lewes and Brighthelmston* (Lewes, 1795),pp. 536–7.

36. C. Parkin, *The history and antiquities of Yarmouth* (London, 1776), p. 400; Anon, *An historical guide to Great Yarmouth, in Norfolk* (Yarmouth, 1806), pp. 20, 61.

37. Evans, *Recreation for the young and the old*, p. 38.

38. O. Bread, *New guide and hand-book to Worthing and its vicinity* (Worthing, 1859), p. 4; Lyons, *Description of the Isle of Thanet*, p. 11.
39. Fanny Burney, *The Early Journals and Letters of Fanny Burney*, ed. LE Troide, (4 vols., Oxford, 1988), I, p. 275.
40. Anthony Relhan, *A Short History of Brighthelmston* (London, 1761), p. 15.
41. *Illustrated journeys of Celia Fiennes*, p. 160.
42. Anon, *The Weymouth Guide* (Weymouth, 1785), p. 57.
43. Relhan, *Brighthelmston*, p. 15.
44. M. Hunter, 'The first seaside house?', *The Georgian Group Journal*, 8 (1998), pp. 135–42, 135–8.
45. C. Miele,' "The First Architect in the World" in Brighton', *Sussex Archaeological Collections*, 136 (1998), pp. 149, 156. This was rebuilt in a grander style by Robert Adam in 1786–7.
46. Anon, *A summer trip to Margate* (British Library, undated [1770s]).
47. Walter Ison, *The Georgian Buildings of Bath*, (2nd edn., Bath, 1980), pp. 5–6.
48. Anon, *Scarbrough Miscellany*, p. 4.
49. R R Angerstein, *R. R. Angerstein's Illustrated Travel Diary 1753–1755*, Torsten and Peter Berg, transl, (London, 2001), pp. 68–9, 227–8.
50. A Brodie et al,, *Weymouth's seaside heritage* (Swindon, 2008), pp. 8–9, 12, 71.
51. Anon, *A companion to the watering and bathing places of England* (London, 1800), p. 21.
52. Steve Peak, *Fishermen of Hastings: 200 years of the Hastings fishing community* (Hastings, 2005).

Chapter 3

A Dutch Idyll? Scheveningen as a Seaside Resort, Fishing Village and Port, c. 1700–1900

JAN HEIN FURNÉE

Nineteenth-century visitors to Scheveningen, the most prominent Dutch seaside resort close to The Hague, did not fail to praise the harmonious co-existence of the village's fishing activities and its expanding leisure industry. In H.W. Mesdag's famous *Scheveningen panorama* (1881, Figure 3.1), we see the large sandy beach neatly divided between a zone for fishing boats and drying sheds and, north of the village, a vast space crowded with bathing machines, cane beach chairs, elegant villas and fashionable hotels.[1] It appeared a classic win–win situation. The picturesque fishing activities contributed to Scheveningen's self-image as a typical Dutch seaside resort, counterbalancing aristocratic luxury and cosmopolitan pleasures with a certain introversion and thriftiness. At the same time, the tourist industry offered the local fishing population a range of supplementary sources of income, from renting rooms and organising donkey rides to selling hand-made souvenirs.

To what extent was this harmonious image of mutual interaction as unproblematic as visual representations and travel guides would suggest? Apart from demonstrating the reciprocal benefits between fishing and resort interests, this chapter will show that this well-cultivated idyll was not free from mutual tensions. In the second half of the nineteenth century, local ship owners increasingly contested the expansion of the resort and various elegant visitors and the resort industry increasingly complained about the smell of drying fish and the rowdy conduct of local fishing families. The growing tensions between fishing and tourist interests became particularly clear in the struggle, which spanned several decades, over the construction of a sea harbour, a process that also involved the residential capital of The Hague.

The Road to Scheveningen

Since the first reference in thirteenth-century official records, the coastal village of Scheveningen has always been dependent on the fishing industry. The late sixteenth-century fish auctioneer, Adriaan Coenen, gave an extensive and lavishly illustrated overview of the village's fishing

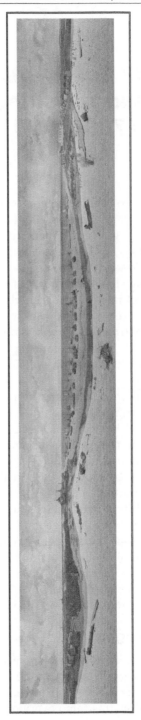

Figure 3.1 Panorama van Scheveningen, H.W. Mesdag, oil on canvas, 1880–1881. From left to right: The canal from the Hague, the village of Scheveningen, the café-restaurant Zeerust (the former bath establishment of Adrien Maas), the local fleet of flat boats, the bathing area, the royal pavilion, the municipal Bath-House and Hotel des Galeries.

activities in his well-known *Vis-Boeck* (Fish book). According to this account, the Scheveningen fishermen sailed out to catch haddock from November until February and codfish and blowfish between February and May. From June, most of them joined the great herring campaigns off the Scottish coast – both in their own flatboats and by signing up large herring boats in Rotterdam, Schiedam and Delfshaven – following the herring to the south until, by the end of the year, they arrived off the Dutch coast. In 1700, the North Sea still possessed a huge natural concentration of fish. Off Katwijk, fishermen caught about 160,000 haddock in a single day. While fresh catches were sold on The Hague's local market, most of the dried plaice from Scheveningen was exported to foreign markets such as Antwerp and even Cologne and Strasbourg.[2] With about 900 inhabitants, most of the 200 households were directly or indirectly involved in the fishing industry, employing about 250 active fishermen and boys alone.[3]

Already in the seventeenth century, a busy traffic existed between Scheveningen and The Hague, not only from fisherwomen visiting the town market but also from inhabitants of the residential capital and foreign visitors taking in the fresh air on the beach. In the 1660s, local government facilitated travel between the two locations by constructing a two-kilometre road straight through the dunes. The initiator, the famous poet and diplomat Constantijn Huygens (1596–1687), intended the road to transform The Hague into a 'fashionable sea town' (*aansienlijcke Zee-stadt*). Consisting of a paved passage for horses and carriages and two footpaths and being protected against drifting sand by substantial dykes planted with trees, it would, in his vision, significantly enhance the transport of fish as well as the influx of 'travelling and leisured persons' looking for some necessary 'refreshing air' (*ververschingh van Locht*). He was not disappointed. The road to Scheveningen greatly stimulated the local fishing industry and transformed the village to a much loved place of recreation, attracting inhabitants from the whole region, foreigners visiting The Hague and tourists from inland provinces such as Gelderland.[4]

For the growing number of visitors to Scheveningen, the proto-sublime experience of looking at the infinite sea went hand-in-hand with enjoying the picturesque qualities of the village's fishing activities. Three years after the completion of the road, the Dutch poet Jacob van der Does described the attractions of the beach in his city praise of The Hague:

> In the morning, when the sun with its hot beams scorches the wanderer
> One can take here a little 'Seaside pleasure' (*Zee-vermaeckjen*).
> And look at the fish auction, and see what fish
> Of our desire or taste has been caught that night (...)
> In the evening, when the Sun dives into the sea
> And closes her all-seeing eye on the world,

One can search here for a little cooling and paddle through the water,
So that the Sea splashes around nose and ears.[5]

This image of Scheveningen as a site of leisure was firmly established
during the next decades. In his *Nederlandsche Historiën* (1679), Pieter Bor
described how The Hague's inhabitants and foreign visitors 'delight and
amuse themselves by watching the sea and eating some fish, arriving
fresh from the sea'.[6] In 1711, Gijsbrecht de Cretscher stressed in his city
encomium that there was hardly a foreigner who visited The Hague for
the first time without 'taking the sea air' in Scheveningen.[7]

Opening the link to Scheveningen greatly benefited the local catering
industry. In 1680, the small village already boasted seven inns, serving
drinks and fish meals, providing spaces to accommodate traditional
sports and organising rich wedding parties from all over the country.[8]
Already in the seventeenth century, the first purpose-built summer
cottage was constructed on the dunes. Surrounded by a beautiful garden,
the owner Cornelis Michelsz Soetens received the Prussian King and
other high-status guests, enjoying the pleasures of watching flat boats
arrive and depart, the fishing activities on the beach and the traditional
costumes of the local fishing population.[9]

Although the Dutch elites clearly discovered the pleasures of the
beach (see Figure 3.2) at a relatively early stage, there is hardly any
evidence that bathing and swimming in the sea belonged to their
repertoire. Only by the end of the eighteenth century did the habit of
sea bathing seem to win some devotees. In 1775, an Italian traveller Pilati
de Fassulo observed that 'many people go to Scheveningen to take a bath
in the sea'. With great delight, he described how he, together with some
Dutch friends, had taken a naked plunge in the sea, assisted by
Scheveningen women and girls who had taken care of their clothes
and helped them to dry off.[10] Nevertheless, in 1806, the Englishman Sir
John Carr remarked in his *Tour through Holland* that the habit of sea
bathing in Scheveningen was still hardly popular. Arriving at the beach –
with 'fishing vessels reclining on the sand in the most picturesque
forms' – he saw how men, women and children on the sand 'paraded up
and down in their Sabbath suits ... as capable of appreciating the beauty
of the scene, as the worshippers of the Steyne at Brighton or of the Parade
of Bath'. However, to his surprise, the Dutch did not follow the English
example in its most essential element: 'Certain it is, that they are not fond
of sea-bathing, otherwise this beach would be crowded with bathing
people, and the country above it with lodging-houses'.[11]

The Establishment of the Seaside Resort

Although Scheveningen cherished its reputation as a place of sublime
wonders and picturesque attractions, the political situation of the

Figure 3.2 The Beach of Scheveningen, Jan Bulthuis after P. C. la Fargue, pen in gray, 1785.

Batavian Republic and the French Occupation (1795–1813) brought the traditional fishing industry into a period of steep decline. As a result of continuous English attacks on the French and their allies, the Scheveningen fishing families lost a substantial part of their fleet. At the same time, from 1806, the French trade blockade on England severely curbed local fishing activities. In order to prevent smuggling, local fishermen were forbidden to sail out by night, and each boat was subjected to extensive and time-consuming inspection. The result for the local economy was telling: between 1795 and 1809, the village's population decreased from 2638 to 2194 inhabitants.[12]

It was in this context of economic decline that the Scheveningen burgher Jacob Pronk Nz. drafted the first proposal for a modern bathing establishment in the village. In 1806/07, he wrote to the Dutch Minister of the Interior that many 'experts' deplored the fact that the Netherlands still lacked any formal seaside resorts. In Scheveningen in particular, he stressed that the need for a dedicated bathing house was pressing. A growing number of local people were getting used to plunging in the sea – both for pleasure and for medical reasons – yet the dangerous gullies in the sand banks caused a significant number of mortalities and made bathing without professional supervision a risky enterprise. At the same time, he remarked, 'many respectable persons refrain from [sea bathing], because their freedom is inhibited by the many spectators and the lack of facilities'. To offer the respectable classes a comfortable and safe passage into the sea, Pronk proposed to provide Scheveningen with a moderate bathing establishment and modern, expensive bathing machines, requesting a 25-year monopoly to return his investments. In his view, the creation of a modern seaside resort would provide an 'outstanding relief for the much impoverished Scheveningen' since local fishing families could compensate for their declining incomes by renting rooms and developing other services.[13]

It took more than 10 years before Pronk realised his plans. Although his project was warmly supported by various Dutch officials, the French authorities appeared to be more interested in promoting inland spas than seaside resorts and clearly did not approve of his strong social ties with leading Orangist families.[14] Shortly after the Restoration in November 1813, the municipal government transmitted his proposal to the local medical committee, who wrote a supportive report.[15] However, it took until 1818 before Pronk finally got permission to start his bathing establishment. Located about half a kilometre north of the head of the village in order not to disturb the local fishing activities, it was a relatively simple wooden building with a sober reception hall and four bath chambers with warm and cold water. Pronk started with only two bathing machines, a small one and a large one, modelled on Ramsgate. A military guard was installed to prevent any disturbances. Patronised by the royal family and favoured by its superb location close to The Hague, contemporaries were sure Scheveningen would, in the near future, reach 'the magnificent level of one of the first and most fashionable seaside resorts of the continental North Sea coast'.[16]

They were not disappointed. With financial support from the local government, Pronk in 1820 transformed his wooden bath-house into a stone building and linked it with the village by a shell-covered path. The total number of baths taken in the summer season expanded from 1400 baths in 1819 to more than 7600 baths in 1826, and the number of bathing machines increased from three to nine. Apart from local inhabitants from The Hague, the seaside resort quickly attracted English and German visitors, including princely families from the highest ranks, such as William of Prussia (the nephew of the Dutch Queen).[17] However, this evident success also posed its own problems. In the early 1820s, the rooms and catering facilities in the village were quite spartan, and Pronk's simple bath-house was increasingly regarded as 'not quite sufficient for distinguished personalities'. In 1824, the mayor of The Hague, after a formal study trip to Boulogne-sur-mer, proposed to build a brand new municipal bath-house that would be superior to 'all those at present erected for this purpose'. Explicitly directed at the highest social circles on the continent, he justified the investment as fundamental for the prosperity not only of the village of Scheveningen but also for the residential city of The Hague. The new bath-house, completed in 1828 and replacing Pronk's, exceeded all expectations. The central building, adorned with a fashionable colonnade, housed a brilliantly decorated dinner room, a billiard room, and a library as well as luxurious suites for princely and other highly distinguished families. Next to two circular wing buildings with guest rooms, the two flanking corner buildings accommodated separate bathing establishments for men and women. In front of the complex, a vast terrace was created with a comfortable entrance to the beach.[18]

In the 1830s and 1840s, the elegant municipal bath-house (see Figure 3.3) definitively turned Scheveningen into one of the most fashionable seaside resorts of the continental North Sea coast, attracting a steady flow of aristocratic families from Germany, Austria and Russia, including foreign princes and even the kings of Bayern and Würtemberg.[19] In accordance with the Dutch cult of polite simplicity, the officially appointed bath doctor L.F. d'Aumerie praised the 'pleasant and convivial atmosphere' that characterised the social life, suggesting that during music evenings and special dinners 'the distinction between the classes was completely forgotten and a true bath-freedom and equality reigned'. On sunny days, highly distinguished guests mingled freely with The Hague's

Figure 3.3 The Municipal Bath House of Scheveningen, lithograph in colour, *c.* 1850.

aristocracy and the *corps diplomatique* in 'splendid assemblies' on the seaside terrace. The queen, who regularly visited the adjacent royal pavilion erected in 1826, used to frequent the terrace to enjoy the fashionable crowd 'with the same delight of a mother who resides with her happy children'.[20] In the meantime, the working classes tended to visit the simple coffee and tea houses in the village, while the 'respectable burghers, bakers and confectioners' set themselves on the terrace of Habraken-Logger, an establishment at the head of the village overlooking the beach.[21] At the beach, the number of bathing machines and canvas tents expanded significantly, both in front of the municipal bath-house and increasingly in front of the village, where in 1844, Logger's successor Adrien Maas started a bathing establishment by the sea. Between 1834 and 1850, the total number of baths recorded increased from about 13,000 to 23,000.[22] According to an informative guide for visitors, the beach showed a fascinating mix of people: sick and healthy, thin and fat, and brown and blond 'all together and apart in the most happy, paradise-like innocence, communistic equality and republican freedom'.[23]

The picturesque quality of Scheveningen as a fishing village was one of its great touristic attractions. For various contemporaries, it was the combination of the sublime spectacle of the endless ocean and the picturesque sailing boats that made contemplating the sea such a pleasure.[24] According to Dr d'Aumerie, looking at fishing activities even had positive medical effects: 'The sick can forget themselves in contemplating the great works of Nature as well as industrious human businesses'.[25] Observing fishing families walking in their traditional national costumes offered 'a tableau unique of its sort' and created a joyful experience by a temporary transgression of social borders. Buying fish just caught in the sea was marketed as one of Scheveningen's greatest attractions, as also was looking for souvenirs: 'Here and there one sees behind the glazed windows a variety of shells and horns, manufactured in different ways and in different shapes as dolls, ships, and even wind mills with eelgrass'.[26] The high value put on the picturesque qualities of the traditional fishing activities had a clear ideological dimension. They offered a counterbalance and, in a way, 'protected' the typical Dutch national identity based on diligence and thrift against the morally corrupting influence of foreign, particularly French, elite bathing culture centred on aristocratic luxury and cosmopolitan pleasures.

Obviously, the local fishing population benefited from the expansion of the seaside resort as well. After the Batavian–French period, the fishing industry only slowly recovered from its decline. In order to stimulate the international market, the new national government in 1814 concentrated the lucrative fish trade in salted herring in three towns on the river Maas (Maassluis, Vlaardingen and Schiedam) and prohibited fishermen from Scheveningen and other coastal villages joining the annual visits to the Scottish coast. Several schemes to build new flat boats to stimulate shore fishing failed. Although the number of Scheveningen flat boats almost doubled from 60 in 1816 to 100 in 1840, the quantity of haddock and codfish in the coast area steeply declined.[27] For the growing population of Scheveningen (4000 people in 1825 to 6000 people in 1850), renting rooms, selling fresh fish and souvenirs and offering services such as donkey rides and sailing trips offered a more than welcome supplement to their incomes.[28]

Growing Tensions

In the second half of the nineteenth century, the fate of the Scheveningen fishing industry turned dramatically. In response to the rise of liberalism, in 1857 the government passed new legislation that cancelled all restrictive measures and allowed the ship owners and fishermen along the North Sea coast the freedom to sail out when, where and how they wanted and to catch what they desired.[29] Interestingly, it was the successful owner of the bath establishment in front of the village, Adrien Maas, who was the first to fully recognise the potential of the new law. Expanding his business as a ship owner, he exchanged the traditional obligatory heavy hemp fishing net for a lighter industrial one made of cotton, which doubled and even trebled the average herring catch per boat. In the mid 1860s, he also introduced a new type of ship, the 'logger', which was much lighter, faster, steadier and more spacious than the traditional flat boats. As most Scheveningen ship owners followed his example, the total income of the local herring catch expanded dramatically from a mere *fl.* 127,000 in 1851 to *fl.* 337,000 in 1860, *fl.* 677,000 in 1871 and *fl.* 2.2 million in 1890. The size of the local fishing fleet extended as well: from 100 in 1850 to 323 in 1900, more than half of the total Dutch fishing fleet. Since the modern loggers with a keel could not land on the beach, the majority of the Scheveningen fleet sailed on to Vlaardingen and Maassluis. Yet Scheveningen also retained a large number of flat boats, operating daily along the coastal strip.[30]

In the same decades, the seaside resort also experienced a spectacular expansion. In order to attract more visitors and compete with the upcoming resort town of Oostende, the local government in 1855 expanded the municipal bath-house with two extra wings, adding in 1866 two further adjacent buildings housing a Kursaal and a vast dining room. In the meantime, a growing number of private investors entered the market. In 1858, a small group of prestigious figures from The Hague joined together in a limited liability company to establish the Hotel Garni, north of the bath-house. Stimulated by the gondola service from The Hague, introduced in 1860, and the horse tram introduced in 1863, speculators started to build a succession of elegant villas in the dunes. From the 1870s, a building mania set in, starting with the construction of Hotel Orange (1874) next to the municipal bath-house, Hotel des Galeries (1876) just behind them and a skating rink (1879), modelled after the English fashion. In the early 1880s, after much debate on the role of the local government as owner and exploiter of leisure facilities, a conglomerate of German and Dutch investors finally bought the municipal bath-house and erected in its place the magnificent Kurhaus, opened in 1885. As the number of foreign and Dutch holiday guests and day visitors grew significantly, the number of recorded sea bathings expanded from 19,800 in 1856 to over 65,000 in 1880.[31]

As the resort and fishing industry expanded dramatically in these decades, their mutual relations also intensified. Making his fortune through both a renowned bathing establishment and a new and expanding shipping company, Adrien Maas is a unique example of how both industries were sometimes integrated in an extremely successful way. On a smaller scale, local fishing families continued to make an increasing income out of the growing number of tourists by selling fresh fish and souvenirs and offering all kinds of leisure services. At the same time, the significance of the traditional fishing character of the village for the branding of Scheveningen as a seaside resort expanded as well, not the least by the immense popularity of the grey-toned beach and sea paintings of local artists such as H.W. Mesdag, J. Israels and J.H. Weissenbruch, since the mid-1870s known as The Hague School. In their paintings of Scheveningen beach, the still much-contested cosmopolitan and increasingly commercial character of the modern seaside resort is hardly visible and completely overshadowed by the trustful and steady character of hard-working local fishing families. Although travel guides praised the increasing modern comforts of the resort, the busy fishing activities at the beach stayed one of Scheveningen's special selling points. It was claimed that at the family hotel Rauch at the village's head 'one can contemplate and learn to know the true fisher's life much better than in the Hotel Garni or the Bath house'.[32]

Despite their mutual interests, the double expansion of both Scheveningen's fishing industry and the resort inevitably fuelled growing mutual tensions as well. During the late 1850s and early 1860s, the use of the beach space at the head of the village was increasingly contested. Maas, who had been running a successful bath establishment (see Figure 3.4) for 10 years, in 1854 received formal permission from the local authorities to use a strip 280 metres wide on the beach to locate his bathing machines, chairs and bathing clientele. A few years later, the local government expanded his exclusive bathing zone to 400 metres and also granted a competing bath entrepreneur a strip of some 200 metres. For the local ship owners and fishermen, this exclusive zoning policy

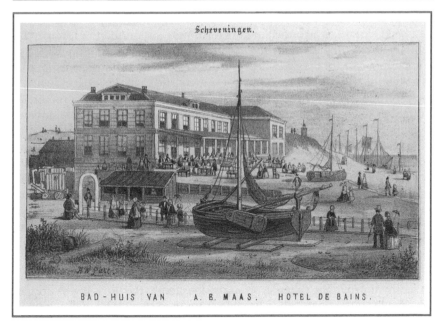

Figure 3.4 The bath establishment of A.E. Maas, H.W. Last, lithograph in colour, *c.* 1855.

seriously harmed their businesses. Not only did it hinder the arrival of their ships and the transport of fish to the village, but the increasing number of flat boats lying next to each other during the night on a decreasing part of the beach also ran the risk of being badly damaged by heavy weather. In June 1861, the Scheveningen ship owners and fishermen with over 400 signatures successfully petitioned against Maas, forcing the local authorities to reduce the beach area designated for bathing to a mere 120 metres and – more important – to declare that the future expansion of the seaside resort should be concentrated in the northern area, leaving the southern area for the expansion of the local fishing industry.[33]

Although the Scheveningen ship owners initially closed ranks against Maas, in the mid-1860s, they joined the successful entrepreneur in his critique of the city government of The Hague, which seemed to favour the resort interests above those of the growing fishing industry. In 1864, the English owner of the local gas factory Neville Goldsmid, after becoming the main concessionary of the new horse tramway to Scheveningen, informed the council that he had bought the former club house of the ship-owners Zeemanshoop and planned to transform it into a magnificent new hotel. To the council's surprise, he proposed to acquire the municipal bath-house as well, arguing this would convince Dutch and foreign capitalists to join the limited liability company he planned to create. Maas, who in 1863 had been elected as a council member, strongly supported the commercial request. According to him, the local government should stop its 'false competition' with the private resort investors and transfer its public investments to provide the infrastructure for the rising fishing industry.[34]

Although sympathetic towards Maas's support for commercial initiatives to expand the seaside resort in order to transfer public money to the local fishing industry, Scheveningen ship owners in the succeeding decades did not hesitate to obstruct the commercial expansion of the seaside resort. This was particularly clear with the multiple initiatives to provide Scheveningen with a modern pier. The phenomenon of the pier, as is well known, had been popular in the English resorts from the 1810s. In 1868, an English consortium chose Scheveningen as the first continental resort on the North Sea to have a modern, purpose-built walking pier, adorned with stalls, restaurants, and a range of leisure facilities. Their request to the local government met initial support from The Hague's mayor and aldermen, yet the board of the local fishing industry, being asked to approve the scheme, strongly objected to the idea. It was argued that the local flat boats when loading from the Scheveningen beach often drifted hundreds of metres to the north before they crossed the surf. The construction of a pier in front of the municipal bath-house would, in their view, seriously run the risk of the boats colliding with it and being damaged. The original English promoters were discouraged by this response, and subsequent initiatives from English and Amsterdam entrepreneurs in 1873, 1876 and 1880 were also obstructed, meeting the same fierce resistance from the Scheveningen ship owners, who organised petitions and successfully convinced the local Chamber of Commerce that the pier would be a deathblow for the fishing industry.[35]

As the leading ship owners increasingly contested and, to some extent, successfully contained the expansion of the seaside resort, a growing number of resort guests, although celebrating the purity of the fishing village, started to question the conduct of the local fishing families. Already in 1850, Maas had acknowledged that the fish-drying sheds were a picturesque sight but at the same time spread a much less pleasant smell. In the 1880s, a chronicler of the social life in Scheveningen complained that 'the "ignoble" dab fish under the eyes of the resort guests spoiled the complete area'. The noisy presence of the Scheveningen fishing families also created a growing nuisance.[36] When in 1885 the new Kurhaus constructed a highly controversial iron fence around the fashionable seaside terrace, some chroniclers saw this primarily as a means to exclude the 'real vulgar Scheveningen ... the men with their caps on the back of their heads, hands in their wide pockets and pipes in their mouth, the women with their wiggling walk, the young ones eating the sugar and licking the empty tea and coffee cups'. Although some contemporaries criticised the exclusion of the Scheveningen families as a loss to the peculiar character of the resort, others appeared happy that these ordinary folk no longer 'offend the eyes and the noses of the superior race of the respectable people'.[37]

The Harbour Question

In the course of the nineteenth century, the relationship between the resort and fishing interests in Scheveningen was further complicated by one of the most contested and protracted planning projects in the history of Scheveningen and The Hague: the construction of the sea harbour. Already in 1824, mayor Copes van Cattenburgh had developed a serious plan to construct a sizeable harbour between the village and the newly projected municipal bath-house. This sea harbour, designed for the packet boat service to Yarmouth, would, in his view, stimulate both the local resort and fishing interests: it would be aimed explicitly to lure English visitors from Oostende and Calais to Scheveningen and would at the same time offer a safe haven for the local fleet of flat boats. The project was warmly supported by King William I, but surprisingly the local ship-owners, intended as co-financiers of the project, were hardly enthusiastic. In the same vein as their later protests against the piers, they argued that the two stone pier heads of the harbour would endanger their flat boats when drifting along the coast. In 1830, the drawings finally disappeared inside the proverbial drawer.[38]

However, in 1852, the idea of a sea harbour reappeared on the local political agenda, to stay there for more than half a century. Although in general not very fond of investing in the city's productive economy, the local council reconsidered the construction of a sea harbour with a direct link to England and a future connection to the quickly expanding European railway network. The council saw it as providing a positive stimulus to the local economy and Dutch industry in general, which had made a very poor impression at London's Great Exhibition. In contrast to the 1820s, the council now intended to locate the harbour south of the village, close to the fish processing industry and intentionally far away from the expanding bathing establishments, which lay between the village and the municipal bath-house. A number of specific plans received warm support from both the local newspapers and the Chamber of Commerce. The ship owners from Scheveningen, however, remained very sceptical, still afraid for their flat boats. When bath entrepreneur Adrien Maas expanded his business as a ship owner, he soon became a main advocate of the harbour as the perfect accommodation for the loggers, the new boat type that he had invented. With his plea that the harbour would transform Scheveningen into one of Europe's largest harbours for fresh fish, he initially won the support of other ship owners. However, when in the 1860s the local council decided that the new harbour should again be located north of the village, the ship owners again objected that it would endanger their flat boats – now extending their argument that the 'strong smell of fish would spoil the pleasures of the resort guests'. In this context of constantly shifting alliances, preferences and projected images of the future and with a local council consisting primarily of fashionable men of leisure who saw The Hague as a city of consumption rather than production, it was no wonder that the six or seven initiatives to construct a harbour in the 1860s and 1870s all failed miserably.[39]

In the mid-1880s, the Scheveningen harbour question turned into an issue in which resort and fishing interests more than ever seemed to clash. When a new committee appointed by King Willem III proposed a sea harbour that would be designed specifically for the new loggers, the majority of ship owners had already exchanged their flatboats for loggers so that the traditional fear of their flat boats running against the projecting harbour heads was replaced by the growing conviction that the harbour would offer a major opportunity for expansion. For the resort industry, now growing substantially in the space between the village and the newly built Kurhaus, the harbour was increasingly seen as a dangerous threat. Suddenly a consensus arose that 'a choice for the harbour is a choice for Scheveningen as a fishing village or a large seaside resort ... One has to choose: fish or meat; in the end they cannot live together'.[40] According to some contemporaries, the seaside resort with its mixed baths and gambling practices increasingly represented an immoral industry. For them, the time had come for the national government to follow the example of Germany and invest heavily in the more productive fishing industry, otherwise in a few years, 'the growing resort here will also have completely pushed aside the fishing village'.[41] Others however argued that the fishing industry offered a growing obstacle to their pleasures, and for a further expansion of the seaside resort, the fishing industry should necessarily be removed from sight and smell:

> If the resort is ever to make something of itself, then there should be a 'divorce of table and bed' ... While a resort guest demands fresh air, a fisherman needs to dry nets and plaice – *incompatibilite d'humeur et de caractere*! There is only one chance to live together in harmony: the *joejen* [i.e. the fishing families] should move southwards and leave this part of the beach to the resort, where they should show themselves solely as a ornament of the landscape.[42]

For many contemporaries, the Scheveningen harbour would not only endanger the seaside resort but the social and cultural character of The Hague as well. If the harbour would indeed transform the city into the 'third commercial port of the country', bustling with energy as the proponents promised, The Hague would no longer be a pleasant, calm residential city for the fashionable leisure classes; 'Then we won't *flaner* any longer!'[43]

Conclusion

In 1904, a few years after a devastating storm had destroyed a substantial part of the remaining fleet of flat boats, Scheveningen finally got its long-awaited sea harbour. Located south of the village, the harbour stimulated a decisive spatial separation between the village's fishing industry and the seaside resort. Although the idyllic image of the harmonious co-existence of the village's fishing activities with the increasing leisure industry was definitively disturbed, this divorce of table and bed literally created the space for both industries to expand into the twentieth century. With the harbour designed for only a small number of fishing boats, Scheveningen would, however, never become the 'third commercial port of the country' as was originally intended. For The Hague's local elites, the personal joys of strolling along the beach, enjoying the terraces and walking the modern leisure pier (1901) were simply too important to run the risk of 'running people, racing merchants, pushing brokers and toiling labourers' taking over the place (see Figure 3.5).[44]

Figure 3.5 'Greetings from Scheveningen', postcard, *c.* 1890.

Notes and References

1. Y. van Eekelen, ed., *Magisch panorama. Panorama Mesdag, een belevenis in ruimte en tijd* (Zwolle, 1996).
2. J .C. Vermaas, *Geschiedenis van Scheveningen* (2 vols, Gravenhage, 1926), II, pp. 1-9.
3. Ibid., I, pp. 23ff.
4. Ibid., I, pp. 144–51.
5. J. van der Does, *'s Graven-Hage met de voornaamste plaetsen en vermaecklijkheden* ('s Gravenhage, 1668) pp. 106ff.
6. Vermaas, *Geschiedenis*, II, p. 388.
7. G. de Cretser, *Beschyvinge van 's Gravenhage behelsende desselfs eerste opkomste, stichtinge en vermakelyke situatie enz* (Amsterdam, 1711), p. 134. For French and English travel descriptions see A. Corbin, *The lure of the sea: the discovery of the seaside in the western world, 1750-1840* (Berkeley and Los Angeles, 1994), pp. 32–9.
8. Vermaas, *Geschiedenis*, II, p. 390.
9. Ibid., I, pp. 115ff.
10. C. Pilato di Tassulo, *Voyage d'Italie et de Hollande* (2 vols, Paris, 1775), II, p. 220.
11. J. Carr, *Tour through Holland along the right and left banks of the Rhine, to the south of Germany, in the summer and autumn of 1806* (London, 1807), pp. 163ff.
12. Vermaas, *Geschiedenis*, I, pp. 237-262, 273.
13. J. Pronk Nz. to Minister van Binnenlandse Zaken en Eredienst van zijne Majesteit den Koning, undated [1806/07]. The Hague, Haags Gemeentearchief (HGA), Library, Ll2-2. Cf. G. Swartendijk Stierling and A. Moll, *Het zeebad of overzigt over den oorsprong en de nuttigheid der baden in het algemeen (...) met een aanhangsel nopens de nieuwe zeebadinrichting te Scheveningen* (Dordrecht, 1819), p. 331.
14. Unknown Dutch official to Minister van Binnenlandse Zaken van het Empire, 25 August 1810. The Hague, National Archive (NA), Archief van het Departementaal Bestuur van Maasland 1807–1810 (toegang 3.02.08), inv.nr. 438; P. H. P. van Marle, 'De Fransche tijd', *Die Haghe* (1904), pp. 135–61, at p. 142f.
15. Minutes Provisioneel Bestuur, 8 July 1814. Rapport Commissie Geneeskundig Toevoorzicht, 30 July 1814. HGA, Oud Archief, inv. nr. 562. Vermaas, *Geschiedenis*, I, pp. 305ff., II, pp. 398ff.
16. Minutes College van B. en W., 27 July 1818. HGA, Archief Stadsbestuur 1816–1851, inv.nr. 32; Swartendijk Stierling and Moll, *Het zee-bad*, p. 335. Cf. Vermaas, *Geschiedenis*, II, pp. 400ff.
17. A. Moll, *Gemeenzame brieven over het Scheveninger zeebad ten nutte van lijders, half-lijders en niet-lijders* (Arnhem, 1824), pp. 21–7. L. F. d'Aumerie, *De zeebadinrigting te Scheveningen en het badseizoen van 1828* (Den Haag, 1829), p. 5. Cf. Vermaas, *Geschiedenis*, II, pp. 402ff.
18. Minutes Stedelijke Raad, 14 May 1824, 12 November 1824, 5 April 1826, 4 February 1828. HGA, Archief Stadsbestuur 1816-1851, inv.nr. 2. *Dagblad van 's Gravenhage*, 15 June 1827; D'Aumerie, *De zeebadinrigting*, pp. 8ff and 116ff. See Vermaas, *Geschiedenis*, II, pp. 406–11.
19. Vermaas, *Geschiedenis*, II, pp. 417.
20. D'Aumerie, *De zeebadinrigting*, p. 6 and pp. 46f.
21. [W. J. A. Jonckbloet], *Physiologie van Den Haag door een Hagenaar* ('s Gravenhage, 1843), pp. 151ff.

22. H. Slechte, *Adrien Eugène Maas (1817–1886). De ziener van Scheveningen* (Scheveningen, 2001) pp. 34ff. *Dagblad*, 7 May 1830, 10 May 1844, 5 June 1850. Cf. Vermaas, *Geschiedenis*, p. 441.
23. Anselmus, *Aan het strand. Lectuur voor badgasten* (Amsterdam, 1849), p. 19.
24. Swartendijk Stierling and Moll, *Het zee-bad*, p. 331.
25. D'Aumerie, *De zeebadinrigting*, p. 20.
26. Anselmus, *Aan het strand*, p. 19.
27. [A.E. Maas], *Wat is Scheveningen en wat zou Scheveningen kunnen zijn?* ('s Gravenhage, 1851) , pp. 3ff. Cf. Vermaas, *Geschiedenis*, II, pp. 27–42.
28. According to the weekly visitor lists in the *Dagblad of 's Gravenhage* in 1850 more than 35 local families earned an extra income with renting rooms.
29. Vermaas, *Geschiedenis*, II, pp. 41ff.
30. Ibid., II, pp. 44ff. See also Slechte, *Adrien Eugène Maas*, pp. 60ff , 98ff.
31. *Verslag over den toestand der gemeente 's Gravenhage over het jaar 1856* ('s Gravenhage, 1857), p. 106; *Verslag over den toestand der gemeente 's Gravenhage over het jaar 1880* (Den Haag, 1881), Appendix 16.
32. 'Haagsche kroniek', *Algemeen Handelsblad [AH]*, 17 June 1876.
33. Slechte, *Adrien Eugène Maas*, pp. 55ff; *Handelingen van den Gemeenteraad van 's Gravenhage*, 13 August 1861.
34. *Handelingen*, 8 and 22 March, 5 April 1864.
35. *Handelingen*, 15 December 1868, 14 October, 18 and 25 November 1873; 27 January, 17 February 1874; 21 March, 30 June, 8 and 15 August, 5 and 17 October, 28 November, 5 and 7 December 1876; 'Brieven uit de hofstad', *Arnhemsche Courant [AC]*, 19 September 1876. In the end, the first modern leisure pier at the continental North Sea coast was constructed in Blankenberghe in 1894.
36. 'Haagsche sprokkelingen', *Utrechtsch Provinciaal en Stedelijk Dagblad [UPSD]*, 29 March 1886.
37. 'Brieven uit de hofstad', *AC*, 2 August 1886. Also *Handelingen*, 7 June 1881.
38. Slechte, *Adrien Eugène Maas*, pp. 122ff; Vermaas, *Geschiedenis*, II, pp. 347ff.
39. Slechte, *Adrien Eugène Maas*, pp. 124-40; Vermaas, *Geschiedenis*, II, pp. 350–7.
40. 'Haagsche sprokkelingen', *UPSD*, 29 March 1886.
41. 'Haagsche sprokkelingen', *UPSD*, 9 May 1887. Cf. ibid., 4 June 1888.
42. 'Haagsche sprokkelingen', *UPSD*, 29 March 1886.
43. 'Haagsche kroniek', *AH*, 7 December 1888.
44. Ibid.

Chapter 4

'From the Temple of Hygeia to the Sordid Devotees of Pluto'. The Hotwell and Bristol: Resort and Port in the Eighteenth Century

DAVID HUSSEY

From the late seventeenth century, the commercial development of the Bristol Hotwell, a site of naturally occurring and widely acclaimed thermal waters, ushered in a sustained period of urban expansion.[1] Founded upon the contemporary vogue for locating therapeutic benefit in the taking of spa water, the Hotwell grew to such an extent that a century later it could boast the full panoply of amenities, distractions and entertainments that was expected of a mature, culturally sophisticated watering place.[2] Complementing the restorative regime, the nearby village of Clifton, Durdham Down and the Avon Gorge offered high-status accommodation and a picturesque, sublime framing to the genteel and polite diversions of the visiting company. The Hotwell thus provided a condensed version of Bath with activities focused upon a clearly defined summer season. However, in addition to the usual paraphernalia associated with spa life – the attendant medical profession, itinerant tradesmen and fashionable shopping, for instance – the Hotwell coexisted alongside arguably the largest and 'most opulent' outport in the early eighteenth century, a major focus of transoceanic commerce and a 'quasi-metropolis' to a sprawling provincial hinterland.[3]

The interaction between the maritime and the cultural is a theme that resonates throughout this volume. Indeed, many early spas and seaside resorts – Scarborough, Yarmouth, Margate, Brighton and the south Wales resorts, for example – existed cheek by jowl alongside substantial port installations that exerted a competing pressure on physical and cultural space.[4] Even at Bath, an inland port of sorts, the improved Avon navigation carved out a commercial zone to the south of the city.[5] However, at the Hotwell, the size and importance of both facilities meant that the occasionally uneasy tension between a resort culture and a port dominated by commercial concerns was amplified. This apparent incongruity was not lost on contemporaries. Although the effusive literature that surrounded the resort served to emphasise the exclusivity of the Company – the oleaginous faux-satire, *Characters at the Hot-Well*, highlights the rarefied, gossipy ambience of the early watering place and its patrons[6] – travellers' opinion of the wider port and city was almost universally disparaging. For example, on his visit to the Hotwell in 1739, Alexander Pope found Bristol to be singularly 'unpleasant' and its mercantile elite uncivilised, a comment echoed in Walpole's famous description of the city as 'the dirtiest great shop'. Less stellar commentators were equally dismissive of the close-built, largely unimproved core of the port, the proximity of manufacturing and industrial enterprises and the 'stinking rill' of the River Avon that brought a throng of shipping and its associated effluvia and detritus into the heart of the city.[7] Although the scale and concentration of trade remained impressive and even disconcerting to observers unaccustomed to the compressed geography of the city, the overriding image was one wherein commerce exerted a rude hegemony over civility and trade leached directly into the cultural forms of urbanity. As late as the 1790s, even the most apologetic of guidebooks could describe Bristol as a 'city long renowned for dirt and commerce' and one that, although partially 'corrected by politeness', remained sullied by its 'disgusting' river and the 'smoke issuing from the brass-works, glass houses etc [that] keeps the town in an almost impenetrable obscurity'.[8]

From this perspective, it was customary to see the Hotwell as an almost epiphytic outgrowth, an oasis of culture and refinement beyond the commercial Gomorrah of the port. To an extent, this has been reflected in the quite discrete historiographies of the port and the spa. However, this apparent oppositional nature translated imperfectly into cultural experience: As has been argued elsewhere, leisure and consumption were central components in a much wider framework of urban development in this period and both Bristol and the Hotwell were no exception.[9] Although it was a fashionable literary trope to view the Hotwell as the very temple of Hygeia, of health and the performance of refinement in contradistinction to a port teeming with the 'sordid devotees of Pluto', these were essentially literary metaphors representing competing modes of production.[10] Viewed through an economic lens, there existed a clear symbiosis between the development of the resort and the expansion of maritime trade. The fame of the resort rested squarely on the imputed restorative values of the well, a chalybeate hot spring that issued some 25 feet above the Avon at low tide and was partially enclosed from inundation and corruption. As at many of the inland spas, the water was considered efficacious as an anti-scorbutic and in the treatment of the usual nephritic and venereal complaints. However, the particular qualities ascribed to the Hotwell in the treatment of both consumptive and diabetic cases quickly ensured its widespread repute.[11]

A key factor at the Hotwell was the early commodification of spa water as an item of conspicuous consumption. The combination of a practically inexhaustible supply, an extensive glass industry and a major port infrastructure led to the development of the trade in bottled water. Table 4.1 outlines the trade in Hotwell water shipped coastally from Bristol to Gloucester, the main entrepot for the industrialising Midlands.

Table 4.1 Hotwell water (baskets) traded coastally from Bristol to Gloucester, 1706–1728

Year		Hotwell water (in baskets)	Shipments
1706		6.00	2
1707		2.00	1
1708		—	—
1709	*	—	—
1710		1.00	1
1711		2.00	1
1712		4.00	1
1713		—	—
1714	*	—	—
1715		3.00	2
1716	*	8.00	2
1717	*	44.00	7
1718		93.00	13
1719	*	40.00	7
1720		165.50	20
1722		405.50	54
1723	*	246.50	42
1724		714.50	83
1725		520.00	76
1726	*	359.75	41
1727		629.25	96
1728		625.00	69

*Data for half year only.
Source: Cox, N.C., Hussey, D.P. and Milne, G.J. (eds) (1997) *The Gloucester Port Books Database, 1575–1765.* Marlborough: Adam Mathew.

Until *c.* 1720, spa water featured mainly as an occasional part-cargo, a useful filler alongside the more established Bristol coastal trades. However, from this date, trade experienced rapid and sustained growth.[12]

In 1724, for example, over 4000 gallons of Hotwell water was dispatched through Gloucester alone, with the commodity being represented in over 42% of coastal voyages clearing Bristol. Even so, this purely localised axis of trade was dwarfed by the extent of metropolitan demand. For example, in 1720, London imported 1944 baskets of water coastwise and in 1725, 3305 baskets were sent to the capital: an unquantifiable, but nonetheless extensive overland trade supplemented these supplies.[13] Hotwell water became an acknowledged 'brand', actively marketed alongside other mineral waters, nostrums and patent medicines, with its legitimacy defended aggressively in the principal newspapers.[14] As contemporary apologists argued, the beneficent effects of the water could be recreated at a distance by the application of indirect heat, which made 'it exert all its virtues in the greater perfection'.[15] Nonetheless, the expedient of revivifying the purity of the water by artificial means could not replicate the full benefit – the 'authenticity' – of taking the waters at the fountainhead.[16] To this extent, a virtuous circle existed between the cultural and commercial arms of the resort, with the lessees of the Hotwell running an efficient bottling operation behind the main Pump Room. Thus, commercialisation of the Hotwell through the marketing of its principal and active component strengthened the commercial reach of the port, which reciprocally promoted the experience of administering the cure directly.

Notwithstanding these apparent mercantile and cultural synergies, the ethos of the resort depended upon maintaining a certain distance: a discreteness wherein the round of health and luxury, of physical and psychological recreation expected by a gentrified clientele could be sealed from both the banality of the quotidian and the often muscular intrusions of the commercial world. Although the distinct cultural and social zoning of other spa developments, notably Bath, served to impart a sense of spatial difference, at the Hotwell, exclusivity was couched in the access to the landscape and, particularly, an extended narrative of the picturesque.[17] Before picturesque tourism became a voguish pursuit amongst the elite, knowledgeable travellers were encouraged to embrace the sublime nature of the Hotwell, to read and digest the painterly magnificence of the natural landscape that was presented to them. Hence, the rhetoric of the picturesque – always an exclusive discourse that sought to restrict bourgeois access – served to demarcate the resort from the port.[18] Key to this was the unusually cramped site of the spa. Situated less than a mile and a half from the crowded centre of Bristol, the Hotwell with its small cluster of amenities and lodging houses was squeezed between the imposing vastness of the 'rough, craggy and romantick' St Vincent's Rock and the River Avon, which wound sinuously, if muddily, towards the city. It was this that most impressed visitors. Even Pope, whose 'thin body' could not bear the exposed setting and reduced level of material comfort out of season, was struck by the main pump house surrounded by 'a Continued Range of Rocks, up to the Clouds, of a hundred Colours, one behind the other'.[19] As the facility improved with the gradual addition of more lodging, a colonnade, piazza and shops, the combination of the mannered, decorous environment of the Pump Room with its orderly company and the rude splendour of the setting was a major factor in ensuring the resort's unique attraction (see Figure 4.1).

The impact of the Hotwell was also closely linked to the development of Clifton, situated some 200 feet above the resort and accessed by a particularly precipitous pathway. A minor rural settlement in 1700, Clifton underwent a period of rapid growth occasioned by the speculative building of residential and lodging accommodation.[20] By 1792, this 'English Montpellier' renowned for the salubriousness of its air had been transformed into the 'largest and one of the most polite' villages in the kingdom and 'a sort of *Westminster* and *Court-end* of the

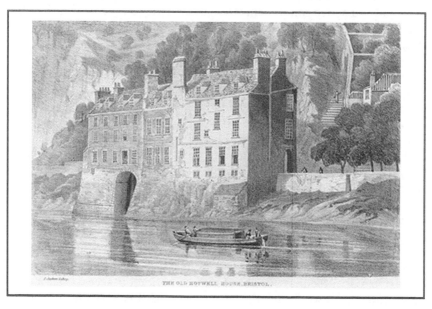

Figure 4.1 *The Old Hotwell House*, Bristol, Samuel Jackson, lithograph, early nineteenth century.

town of *Bristol'* – a focus of both resort activity and quasi-suburban sprawl.[21] Clifton added a stylish *rus in urbe* feel to the resort and supplied refined yet reasonably priced quarters 'genteelly fitted up, and well suited to the purpose of such as come to the *Hotwell* for pleasure only, as well as those who resort thither for the benefit of the waters and the re-establishment of their health'.[22] Moreover, the large tracts of open space from the Downs to Kings Weston offered the visitor, as one commentator enthused, the most 'extensive and glorious' prospects over the countryside that 'the imagination can conceive ... forming a rich display of almost every thing that can render a landscape beautiful'.[23] To a much later tourist, the scenery was 'worthy of the pencil of Salvator and Claude combined, as there are features in it suitable to both those great masters, – the lovely and the terrific'.[24] Beyond such bombast, Clifton permitted the kind of restrained exercise and free association that formed an essential part of the daily regimen of the resort, prescribed for the invalid and valetudinarian alike.[25]

In this context, it is tempting to read the Hotwell as a scenic enclave, culturally if not physically gated by the codes of the picturesque from outside disturbance. This was precisely the impression that the guide-books and quasi-medical tracts wished to confer, and their emphasis on the ordered delights of this 'land of wonders', wrapped up in the increasingly hyperbolic language of the romantic, transferred directly to the poetry of Chatterton and Whitehead and the pages of Burney. In particular, the compressed action of these literary forms mirrored the principal diversions of the Company. With the requirement that partakers should imbibe up to three pints of water each day, activity was necessarily focused upon the public amenities and, in particular, the Hotwell house, Pump Room and the nearby assembly rooms, coffee houses and walkways. Balls, concerts, exhibitions, public breakfasts, promenades, a purpose-built chapel of ease and, for more worldly distraction, luxury shops and a Vauxhall garden located close to the spa absorbed much additional time that was not otherwise occupied by medicinal concerns. By 1766, the new Theatre Royal in Bristol, which had replaced the rather ersatz playhouse in lower Hotwells, added to the available diversions and drew the city and resort into more direct communication.[26]

The polite settings and their associated amusements demanded considerable expenditure in both subscription fees and, indirectly, correct sartorial bearing. As one guest tartly remarked in 1733, the enclosed atmosphere of the Hotwell could be both stultifying and poisonous:

> This is the dullest place that ever was known. There is not above half a dozen families, and those are cits with great fortunes or Irish impertinents: the former despise one because their clothes are finer

than yours; and the latter have no view in keeping your company but to report your faults.[27]

More intrepid visitors could engage in active sightseeing and pleasure seeking. Boat trips along the Avon to King Road, Bristol's deep-water anchorage in the Severn estuary, allowed mixed parties to dine al fresco beyond the regulated torpor of the resort amidst picturesque wooded scenery. Here, in a precursor to later resort practice, 'gentlemen and ladies [can] divert themselves upon a noble Sea beach, in picking up pebbles of all shapes and colours, many of which are so cover'd with particles of mundick, and others with those of isinglass, that they seem full of gold and silver'.[28] Closer to the resort, Rownham ferry, variously described as 'romantic' or 'execrable', took visitors across the Avon to the bucolic charms of the 'sweet and wholesome village of Ashton to eat strawberries or raspberries with cream'.[29] Elsewhere, gentlemen enthusiasts indulged themselves in collecting the varied flora, described in full Linnaean precision by the competing guidebooks, or in searching for fossils and the semi-precious 'Bristol stones' – local quartz and spar mined in the Gorge and sold by the bucket-load in the many informal stalls and booths that lined the resort.[30]

In these pursuits, redolent of the nineteenth-century English seaside, it is possible to detect a distinct 'tourist gaze' at play. Constructed and manipulated by a raft of promotional literature, such activities formed both an idealised symbolic space and a visual and experiential itinerary for visitors.[31] Through such means, the landscape, carefully sanitised of its commercial underpinnings, was systematically presented for consumption. The Bristol stones, blasted out of the rock by faceless quarrymen, mute figures in a socially stable, picturesque landscape, and hawked as souvenir trinkets by poor women, were redeployed as decorative mementos in the establishments of the visiting elite. Indeed, the renowned Clifton garden and grotto of the Bristol merchant Thomas Goldney, a must-see for the beau monde, formed a clear template for those taking the waters.[32] The palpable sense of 'otherness' derived from the setting also filtered into the representations of the port. Normal maritime activities – the procession of coastal and overseas craft using the Avon, the workings of the docks, the loading and unloading of exotic cargo and the ballast grounds, for example – were reduced to scenic backdrops upon which the performance of labour was acted out for the delight of the visitor. The combination of the port as a voyeuristic spectacle and trade as a reified tourist experience was clearly exploited by the guides to provide virtual journeys into the romantic. For example, George Manby in perhaps one of the most effusive passages of the genre recounts the 'enchanting sight' of two West Indiamen being towed out of port from the vantage point of St Vincent's Rock. In the wake of the merchant vessels, a human flotsam of tearful wives and bereft children follow on the towpath. As a true man of sentiment, the leitmotiv of refined masculine performance, Manby re-imagines plebeian experience for tourist gratification: 'to see women in distress', he remarks, 'what human heart can behold them, and remain unmoved'.[33]

The distractions of the Hotwell may well have been distanced from the more insular cultural horizons of the city, and it is important to remember how far civic codes were conditioned by a sense of wariness towards the rural other throughout much of the eighteenth century.[34] Even so, in the early years of the resort, a clear gap existed between the seasonal influx of visitors and the availability of suitable lodgings. The Pump Room and associated buildings permitted only limited on-site accommodation.[35] Clearly, the city had to absorb the seasonal overspill at least until the concerted building of more appropriate dwellings began, firstly with the Dowry Square and Parade development in the 1720s and thence by waves of speculative construction into Clifton. Writing in 1703, John Underhill, in an extraordinarily overblown piece of puffery, attested to the fact that many early visitors resided in College Green, one of the more respectable and, as Underhill – the resident physician for hire – impressed on his audience, hygienic districts of the town.[36] The city nevertheless remained an alternative base for those less enamoured by the especial delights and showy ambience of the Company.

The pressure of demand on the Hotwell and its relatively stretched facilities was marked in the peak years between 1760 and 1790: in the summer season of 1776 alone, 'well over' 700 persons of quality came to the resort, and in the following decade, attendances of over 1000 such grandees were not uncommon.[37] Of course, such figures do not account for the entourage and servants of visiting worthies or the increasing numbers of individuals of lesser status frequenting the resort. Bourgeois visitors and, in particular, genuine invalids were hardly the sort to feature prominently in society lists. In the manner of John Wesley, who in 1754 holed himself up in the isolated and inaccessible New Hotwell, a practically redundant minor adjunct to the main site, many of the aspirant middling sort were unwilling or, more pertinently, unable to avail themselves of the round of subscription-laden distractions and the heavy cultural capital that underpinned them. For example, when the Lancaster merchant and Quaker Dodshon Foster visited the Hotwell in 1766, he, his consumptive wife and their maidservant took rooms in the Rock House, almost directly opposite the Pump Room. The rest of this family, however, found cheaper accommodation at the White Hart in Bristol. Apart from attending the Pump Room, the Fosters preferred a more private experience, wherein dining alone or with select friends was punctuated by riding and occasional sightseeing – Goldney's subterranean grotto was a prominent if damp excursion. Indeed, whilst his wife was recuperating, Foster spent much of his time visiting relations and co-religionists at the resort and in Clifton, being drawn into Bristol to attend the Quaker meeting house, to conduct business or to peruse the comings and goings at the port in solitude. Shopping for the require-ments of polite entertaining – general provisions, tea, coffee and chocolate – also appears to have taken place in the city.[38] In these ways, port and resort were brought together, although for religious and practical reasons; Foster's engagement with spa culture was conditioned almost wholly by the health of his wife and the dictates of the Hotwell regimen. There is no sense here that Foster engaged in the wider social diversions of the assembly rooms and coffee houses that were a marked feature of the resort.

Whilst Foster's low-key encounter with the Hotwell was probably quite a common experience for many middling sort visitors to the spa, it also indicates that as the Hotwell grew in popularity and the Company more heterogeneous, the cultural framework of taking the waters became a more displaced and singular practice. By the 1770s, the occasional, passing tourist had emerged: enterprising owners were advertising rooms convenient for 'parties who would wish to stay ... for a few days', and common lodging houses in lower Hotwells, 'generally filled with people of various kinds – some sick, some well, some dying, some dancing', provided cheap middle-rank accommodation.[39] In addition, the exclusivity of the resort was further diluted by the presence of many of Bristol's own resident bourgeois,[40] and as landlords, the Society of Merchant Venturers jealously reserved the right for members, their families and associates to frequent the Pump Room and use the facilities gratis.[41] Within these contexts, the appointment of a master of ceremo-nies appears less of an attempt to ape the practice of Bath than a device to corral what was fast becoming a fragmentary, disparate activity.[42] Yet, beneath the broadening spectrum of parvenu visitors and nouveaux riches lurked the spectre of democratisation or, more simply, plebeian intrusion. By customary practice, the citizens of Bristol had free access to the original well-head, a right defended robustly by the Common Council. However, from the mid-eighteenth century, there were dark mutterings that '[e]very fine Sunday' the Hotwell was 'all day long like a fair', with the consequent sense of disorderly and unregulated behaviour occasioned by the 'vast numbers of people coming from Bristol, and all round the country, to drink the water'.[43] Although such cultural interlopers were carefully screened – admitted through 'a back way' so as not to 'interrupt the better sort of company' – there are suggestions that the cordon sanitaire that the proprietors and Company fondly imagined to seal the resort of health and exclusivity was far more permeable than they would willingly countenance. As such, the Hotwell was probably far more of a contested space than contemporary eulogies to gentility and order often allow.[44]

These innate tensions remained largely masked by the continued attraction of the Hotwell and the belief in the medical efficacy of its waters. However, just as the port of Bristol experienced profound dislocation to its traditional transoceanic staples in the later eighteenth century, so the Hotwell came under pressure from falling demand and changes in the cultural process of taking the waters. In 1785, the 90-year lease of the Hotwell reverted to the Society of Merchant Venturers, thereby concentrating direct control of both port and resort in the hands of Bristol's mercantile oligarchy. The Society, keen to consolidate future revenue after patently underestimating the resort's commercial potential in 1695, sought a willing tenant with deep pockets.[45] Nonetheless, subsequent surveys revealed a state of general negligence and structural 'dilapidation'. The Pump Room was functional but spartan: 'A most irregular, un-meaning compound, jumbled together so as to make it impossible to bring it to any reasonable form by adding or repairing' as one surveyor tartly put it.[46] Furthermore, the site lacked the necessary refinement for polite entertainment on the scale of its rivals; its cold and damp facilities 'incommoded' the company, who had to pick their way past an ad hoc assemblage of public houses, inns, lodging houses and the semi-permanent shops of seasonal traders to reach the well.[47] More ominously, the corruption of the source from the inundation of polluted tide and river water threatened to disrupt the therapeutic regimen upon which the resort depended.[48]

Despite circulating the details of the Hotwell lease, the Society's demands – £1500 expenditure on material improvement and a punitively expensive repairing lease thereafter – proved singularly unattractive.[49] However, the correspondence with prospective developers reveals how the merchants visualised resort space. Primarily, the Society wished to project 'a respectable figure as well to the navigable world as to those in the water drinking season, [who] resort thither for the sake of health or pleasure'. However, notwithstanding the fondly held desire to make the resort 'the admiration of all Europe', the bottom line remained finance: any remedial work was predicated upon interim financial considerations.[50] As a result, the Society rejected a comprehensive programme of modernisation favouring piecemeal improvement. This provided an elegant new colonnade with fixed shops, a covered way, piazza, much-needed sea defences and a rather superficial refurbishment to the interior of the Hotwell house, but it failed to fully address the core problem of water supply (see Figure 4.2).

The Society's motives appear to be defined by the economics of mercantile exchange, and their dealings lacked the flexibility and cultural adroitness required in directing an elite resort. Increased subscription

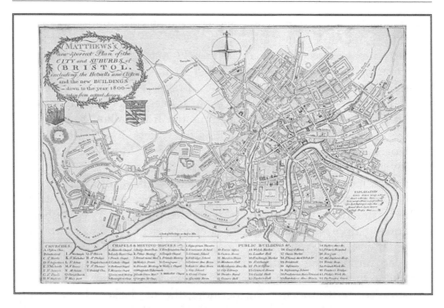

Figure 4.2 Matthew's New and Correct Plan of Bristol, 1800.

fees and a desire to police their collection more strictly may well have translated into improved revenue streams in the short term but conspired to dissuade the more casual tourist over time. According to one contemporary, the charge of 26s per month for an individual to drink the waters, levied in 1790, served only to limit access for the desperate and incurable: The Hotwell had become 'practically a fountain sealed to the lips of everyone but the actually moribund'.[51] Indeed, as Figure 4.3 demonstrates, annual proceeds from the Pump Room, its associated amenities and the wider sale of water declined sharply as the period progressed.[52] Certainly, the impression of the crowded, fashionable spa of the 1780s and 1790s was swiftly displaced by one that stressed the decay and tawdriness of the resort. In the words of one critic writing in 1816, 'one of the best frequented and most crowded watering-places in the kingdom' in 1789 had been reduced to a place that had 'the silence of the grave, to which it seems the inlet'.[53] Six years later when the Hotwell house was closed for a belated refit, the resort was barely breaking even,[54] and by 1824, steam packets were plying their trade to the Devon resorts from nearby quays.[55]

In addition to cost, the progressive erosion of the exclusive cultural space surrounding the Hotwell was a key issue in the decline of the resort. The resort suffered tangentially from the collapse of the speculative building boom in the 1790s that left bankruptcy and recession in its wake and much of Clifton blighted by the skeletal remains of

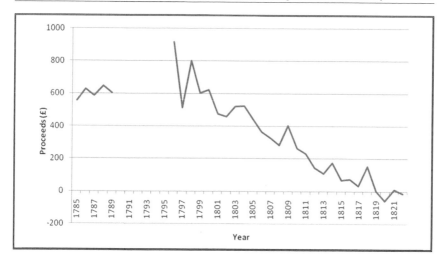

Figure 4.3 Annual proceeds from the Hotwell, Bristol, 1785–1822.

half-finished projects.[56] In a similar vein, the pressing requirements of long-overdue dock improvement, culminating in the creation of a floating harbour in 1809, extended wharfage with its attendant ancillary trades into the immediate environs of the Hotwell.[57] In 1799, travellers visiting the 'calm' of Hotwell had to endure 'running through a kind of gauntlet' of commercial and maritime Bristol; after 1809, the gauntlet extended to the very approaches of the spa.[58]

Underpinning these factors was the steady drift of summer visitors to more fashionable inland and seaside resorts. Part of the resort's problem was its limited season – 'the real cause of so frequent failures amongst the Lodgeing House Keepers'.[59] With one eye on the success of the resorts of Devon, Cornwall and Wales,[60] attempts were made to establish the Hotwell as a superior urban 'winter retreat for consumptive invalids', possessing 'all the advantages of a maritime situation, without the disadvantages; the mildness of sea air, with an exemption from storms and tempests'.[61] Whilst such claims promoted some over-wintering, they did little to dispel the pervasive image of the Hotwell as more of a sanatorium for 'wretched spectres' awaiting their inevitable quietus, than a seat of pleasure, frivolous expenditure and genteel gaiety.[62] In this context, the example of James Oakes, a wealthy Bury St Edmunds banker, is instructive. In the winter of 1797–1798, Oakes trekked across the country to seek a last desperate cure for his terminally consumptive daughter.[63] Tellingly, Oakes was an avid frequenter of the most prominent watering places and seaside resorts, rarely passing the chance to indulge in vigorous sea bathing or the polite drinking of spa water as the occasion permitted. However, Oakes' experience of the Hotwell was solely in terms of a *'dernier* resort' for the dying.[64]

Throughout the eighteenth century, the port of Bristol and the Hotwell existed in both symbiosis and opposition. The proximity of these two facilities meant that there was a considerable and occasionally uncomfortable physical overlap in what were nominally maritime and cultural spaces. However, whilst both port and resort were concerned with production, whether that was played out in commercial or aesthetic terms, both suffered ultimately from underinvestment and myopic management. The floating harbour undoubtedly prolonged the life of Bristol as a port of national significance, but the transformation of dock space was arguably too late, too expensive and generally unfit for the purposes of increased maritime tonnage. Inflated charges and punitive incidental dues may also have blighted trade in the long run. Similarly, at the Hotwell, the escalation of subscription fees to cover even the modest level of improvement proved a key disincentive to the popularity of the resort, whereas the intrusion of an increasingly foul dock space impaired the exclusivity and ambience demanded by the visiting elite. As a result, both the port and the resort entered the nineteenth century under something of a cloud and although the mercantile economy of Bristol found ways of diversifying its commercial base, decline at the Hotwell was steep and irrevocable.[65]

In many ways, the failure of the Hotwell represented a failure to adapt to the shifting circumstances of resort fashion. Although other spas and seaside resorts changed emphasis – for example, at Scarborough where sea bathing displaced the earlier focus of taking spa water or simply relocated to accommodate port expansion, as Louise Miskell's chapter on Swansea indicates – the Hotwell singularly failed to develop alternative sites or practices. Always a poor cousin to Bath, the Hotwell lost out to both the newly fashionable and elegantly constructed Cheltenham and the shift in emphasis towards sea bathing; later attempts to revive resort culture and hydropathic treatment at both the rebuilt Hotwell and at Clifton proved stillborn. By 1841, whilst Clifton still retained the faded aura of a modish spa, it had become a retirement 'colony of half-pay notables, who have lineage and little cash' and a few invalids. The Hotwell, however, had 'ceased to attract'; its 'semi-tepid sparkling water' was shunned by the medical community and fit only for the curious tourist and the occasional passing seafarer, the sole 'admirer of the fair Naiad'.[66]

Notes and References

1. otes:[1] V. Waite, 'The Bristol Hotwell' in P. McGrath, ed., *Bristol in the eighteenth century* (Newton Abbot, 1972), pp. 109–26; B. Little, 'The Gloucestershire spas: an eighteenth-century parallel' in P. McGrath and J. Cannon, eds., *Essays in*

Bristol and Gloucestershire history (Bristol, 1976), pp. 170–99; P. Hembry, *The English spa, 1560–1815* (London, 1990), pp. 245–50.

2. P. Borsay, 'Health and leisure resorts, 1700–1840' in P. Clark, ed., *The Cambridge urban history of Britain. Volume II: 1540–1840* (Cambridge, 2000), pp. 775–804; J. Towner, *An historical geography of recreation and tourism, 1540–1940* (Chichester, 1996), pp. 54–64; J. Stobart, 'In search of a leisure hierarchy: English spa towns and their place in the eighteenth-century urban system' in P. Borsay, G. Hirschfelder, and R-E. Mohrmann, eds., *New directions in urban history* (Munster, 2000), pp. 19–40.

3. See K.O. Morgan, *Bristol and the Atlantic trade in the eighteenth century* (Cambridge, 1993); D.P. Hussey, *Coastal and river trade in pre-industrial England* (Exeter, 2000).

4. See the chapters by Brodie and Miskell in this volume.

5. See R. S. Neale, *Bath, 1650-1850. A social history* (London, 1981), pp. 117–19, 202–3.

6. Anon, *Characters at the Hot-Well, Bristol, in September, 1723* (London, 1724).

7. P.T. Marcy, 'Eighteenth century views of Bristol and Bristolians' in McGrath, *Bristol*, pp. 11–40.

8. E. D. Clarke, *A tour through the south of England, Wales and part of Ireland* (London, 1791), pp. 148–9 and J. C. Ibbotson, *A picturesque guide to Bath, Bristol Hot-Wells, the river Avon and the adjacent country* (London, 1793), pp. 155–6, 188.

9. See P. Borsay, *The English urban renaissance* (Oxford, 1989); J. Stobart, et al., *Spaces of consumption* (London, 2007), pp. 57–85; and C. B. Estabrook, *Urbane and rustic England* (Manchester, 1998), pp. 1–19.

10. R. J. Thorne, *Bristolia, a poem* (Bristol, 1794), pp. 12–13; W. Whitehead, *An hymn to the nymph of Bristol spring* (London, 1751), pp. 11–12; and Marcy, 'Bristol', p. 38.

11. For a survey of tracts see G. Randolph, *An enquiry into the medicinal virtues of Bristol-water* (London, 1750), pp. 1–32 and John Nott, *Of the Hotwell waters* (Bristol, 1793), pp. 10–21.

12. Hussey, *Coastal trade*, pp. 82–4.

13. S. McIntyre, 'The mineral water trade in the eighteenth century', *Transport History*, 2 (1973), pp. 1–19, at pp. 4–5, 12–13; Hussey, *Coastal trade*, pp. 15–16 for conversions and discussion.

14. *Mist's Weekly Journal*, 30 Apr. 1726; Bishop and Battersbee in *Daily Journal*, 11 Aug. 1729; 22 Jan. 1730; and Thomas Woodall: *Gazetteer and London Daily Advertiser*, 17 Nov. 1762; *London Evening Post*, 18 May 1773.

15. E. Owen, *Observations on the earths, rocks, stones and minerals about Bristol* (London, 1754), pp. 132–3.

16. Nott, *Hotwell waters*, p. 87.

17. Borsay, 'Health and leisure resorts', pp. 792-3; J. K. Walton, *The English seaside resort* (Leicester, 1983), pp. 6–10; and Neale, *Bath*, pp. 12–48.

18. M. Andrews, *The search for the picturesque* (Aldershot, 1989), pp. 39–66.

19. G. Sherburn, ed., *The correspondence of Alexander Pope* (5 vols., Oxford, 1956), IV, pp. 201–5.

20. Owen, *Observations*, p. 126; Shiercliff, *The Bristol and Hotwell guide* (Bristol, 1793), pp. 76–9; J. R. Ward, 'Speculative building at Bristol and Clifton, 1783–1793', *Business History*, 20 (1978), pp. 3–18.

21. See Ibbotson, *Picturesque guide*, pp. 174–5; G. Heath, *The new history, survey and description of the city and suburbs of Bristol* (Bristol, 1794), pp. 107–8.

22. Shiercliff, *Bristol*, p. 76.

23. Clarke, *Tour*, pp. 151–2.
24. A. B. Granville, *The spas of England, and principal sea-bathing places. Southern spas* (London, 1841), p. 354.
25. Nott, *Hotwell waters*, pp. 83–97.
26. See Stobart, 'Leisure hierarchy', p. 29; J. Barry, 'The cultural life of Bristol, 1640-1775', D. Phil. thesis, University of Oxford, 1983, pp. 187–90.
27. Miss Kelly to Swift in W. Scott, ed., *The works of Jonathan Swift* (19 vols., Edinburgh, 1814), XVIII, pp. 186–8.
28. Owen, *Observations*, p. 12.
29. Heath, *New history*, p. 104; A. Carrick, *Dissertation on the chemical and medical properties of the Bristol Hotwell water* (Bristol, 1797), pp. 81–2.
30. See Nott, *Hotwell waters*, pp. 25–41; Shiercliff, *Bristol*, pp. 82–97; J. Chilcott, *New guide to Bristol, Clifton and the Hotwells* (Bristol, 1826), pp. 196–7.
31. J. Urry, *The tourist gaze* (London, 1990), pp. 1–7, 10–13; K. Meetham, *Tourism in global society* (London, 2001), pp. 9–11; 36–9, 81–4.
32. A. Young, *A six weeks tour through the southern counties of England and Wales* (London, 1772), pp. 187–9.
33. G. W. Manby, *The history and beauties of Clifton Hot-Wells* (London, 1806), p. 60. See also, A. Fletcher, *Gender, sex and subordination in England, 1500–1800* (New Haven, 1995), pp. 322–46.
34. Estabrook, *Urbane and rustic England*, pp. 206–7.
35. Bristol Record Office (BRO) SMV/6/6/1/1.
36. John Underhill, *Thermologia Bristoliensis* (Bristol, 1703), pp. 16, 30, 31. See also E. Baigent, 'Economy and society in eighteenth-century English towns: Bristol in the 1770s', in D. Denecke and G. Shaw, eds., *Urban historical geography* (Cambridge, 1988), pp. 109–24.
37. Little, 'Gloucestershire spas', p. 175. *Felix Farley's Bristol Journal* (*FFBJ*) records 1,073, 1,003 and 1,044 visitors in 1787, 1788, and 1789.
38. Lancaster City and Maritime Museum, 1993.38.3, Diary of Dodshon Foster. See also A. E. Hurley, 'A conversation of their own: watering-place correspondence among the bluestockings', *Eighteenth-Century Studies*, 40 (2006), pp. 1–21.
39. *FFBJ*, 14 Apr. 1787: James Barton; *Morning Post*, 23 June 1775.
40. For example, the straight-laced Quaker, Sarah Champion, frequented the Hotwell as occasion and sociability dictated, although she clearly preferred Cheltenham: M. Dresser, ed., *The diary of Sarah Fox* (Bristol Record Society, vol. 55, Bristol, 2003), pp. 21, 28, 49.
41. BRO SMV/6/6/2/3, pp. 3, 7–8.
42. Waite, 'Hotwell', pp. 120–1; Hembry, *English Spa*, p. 247; and Shiercliff, *Bristol*, pp. 74–5.
43. Owen, *Observations*, p. 126.
44. See S. Poole, '"Till our liberties be secure": popular sovereignty and public space in Bristol, 1780–1850', *Urban History*, 26 (1999), pp. 40-54.
45. In 1694, the Hotwell was let for £5 per annum: BRO SMV/6/6/2/1–2.
46. BRO SMV/6/6/3/1: 'Escurial', 24 Mar. 1785.
47. BRO SMV/6/6/3/1: Rosser, 9 Dec. 1784.
48. BRO SMV/2/4/2/33/48: 23 Nov. 1789; SMV/6/6/3/1: Steele and Townson, 5 May 1785.
49. *Whitehall Evening Post*, 17 Mar. 1785; *Morning Post*, 25 Mar. 1785.
50. BRO SMV/6/6/3/1: 'Escurial', 24 Mar. 1785.
51. Dr Carrick quoted in J. Latimer, *The annals of Bristol in the nineteenth century* (Bristol, 1887), p. 72.

52. Data from BRO SMV/6/6/1/1–3; SMV/6/6/3/1. Years run from September. Between 1790 and 1795, the Hotwell was leased out: SMV/6/6/2/3.
53. Latimer, *Bristol*, pp. 71–2.
54. BRO SMV/6/1/2/2; *Bristol Mercury*, 4 January 1819. The Hotwell, rebuilt in the Tuscan style, was not a commercial success.
55. J. F. Travis, *The rise of the Devon seaside resorts, 1750–1900* (Exeter, 1993), pp. 80–3.
56. Ward, 'Speculative building', pp. 13–15.
57. See K. Morgan, 'The economic development of Bristol, 1700-1850' in M. Dresser and P. Ollerenshaw, eds., *The making of modern Bristol* (Tiverton, 1996), pp. 48–75.
58. G. S. Carey, *The Balnea: or an impartial description of all the watering places in England* (London, 1799), p. 145.
59. BRO SMV/6/6/3/1: 'Escurial', 24 Mar. 1785.
60. See Travis, *Devon*, pp. 7–21, 26–47, and the chapters by Borsay and Miskell in this volume.
61. Carrick, *Dissertation*, pp. 77–87; *FFBJ*, 10 Oct. 1789: Hotwells subscription assembly; *World*, 6 Oct. 1790.
62. *Morning Post*, G.G. letter, 23 June 1775.
63. J. Fiske, ed., *The Oakes diaries* (2 vols, Suffolk Record Society, 32-33, 1990-91), I, pp. 357–9.
64. Between 1778 and 1827, Oakes visited Bath, Brighton, Buxton, Cheltenham, Cromer, Exmouth, Lowestoft, Sidmouth, Southwold, Weymouth, and Yarmouth: Fiske, ed., *Oakes*, I, pp. 226–8, 273–4, 280, 315–6, 355–6; II, pp. 74, 102–4, 107–8, 122, 124, 132; 255–6, 283, 300–302. The quote is from J. Chilcott, *New guide*, p. 185.
65. Morgan, 'Economic development', pp. 67–9.
66. Granville, *Spas*, p. 359.

Chapter 5

Three Views of Brighton as Port and Resort

FRED GRAY[1]

Itinerary

This essay describes a short promenade along the seafront of present-day Brighton, England. The perambulation takes in three views. Each view is used to look at contrasting descriptions and explanations – historical and contemporary – of the relationship in Brighton between, on one hand, commercial fishing and other port-related industries and, on the other, the town's resort role. Currently, dominant perspectives provide partial and limited accounts of how the Brighton fishery and resort industries interacted. The argument is made that over a long period, the fishing and seaside holiday businesses reached a sustained and mutually beneficial accommodation and one that was to assist in the enduring resilience of both industries.

View One: Dr Russell's Wall Plaque

A wall plaque on the sea-front Royal Albion hotel in present-day Brighton commemorates Dr Richard Russell with the phrase, 'If you seek his monument, look around'. The Brighton-based Russell, working and writing in the mid-eighteenth century, was the most important publicist for the therapeutic benefits of consuming seawater.[2] The wall plaque suggests that Russell made Brighton.[3] Close by is the evidence of the ensuing two and a half centuries of resort building. It includes the influential 1760s Marlborough House and the early nineteenth-century oriental Royal Pavilion, both on the edge of the Steine (see Figure 5.1), an open space designed for leisure, through to the later seafront piers, promenades and hotels. However, with the exception of a small fibreglass pleasure dinghy marooned on a seafront roundabout, there is no sign of the town's maritime and fishing past. The view of and from Russell's plaque asserts that seaside Brighton is a resort and a resort alone.

Brighton, on England's channel coast and 87 kilometres south of London, was one of the world's earliest modern seaside resorts and the largest and most influential in the West for much of the eighteenth and nineteenth centuries. Unsurprisingly then, and mirroring the sentiment on the Russell plaque, popular and academic accounts also concentrate on this dimension of Brighton's history. Other components of the town's

Figure 5.1 The Old Steine, Brighton, 2008. The merest acknowledgement of the resort's maritime history – a dinghy stranded on a seafront roundabout – with the Steine gardens beyond.

coastal past, and particularly the fishing industry, tend to be ignored, belittled or otherwise brushed aside.

Sometimes the maritime past is presented as an ancient pre-history. One dominant perspective asserts that by the early eighteenth century, a once vibrant fishery had so declined and eroded that Brighton had become an urban vacuum. The emergent seaside businesses of health and pleasure surged forward to fill the empty space and so rescued the town and its people from decay and obscurity. For example, Manning-Sanders, writing of Brighton in the 1730s from the perspective of the mid-twentieth century, described how the place that was 'so soon to blossom into the most fashionable and most famous seaside resort in England … [was then] … a decaying fishing town, if town it could be called'.[4] Inglis, half a century later, chooses Brighton as his prime example of how 'fishing villages became watering places'.[5] The academically most influential adoption of this position is by Shields. He asserts that the development of Brighton beach as a liminal pleasure zone of escape was 'speeded up', during the nineteenth century by the 'decline of the "working beach": that beach which had seen the launching of generations of fishing boats …'.[6]

This view of decline appears to be supported by evidence indicating that until the first decade of the eighteenth century the fishing industry was centred on the Lower Town – a low-lying foreshore protected by an offshore shingle bank. The industry was beach-based, the boats were beach-launched and members of the fishing community lived in the same area and away from the 'higher town' on the low cliffs above the foreshore, which is the location of present-day old town Brighton.[7] Storms destroyed the Lower Town and much of the fishing industry. In frequently quoted passages, Daniel Defoe described the impact of a storm in 1703 as leaving Brighton 'looking as though it had been bombarded' and, two decades later, portrayed the place as 'a poor fishing town, old built, and on the very edge of the sea. The sea is very unkind to this town, and has by its continual encroachments, so gained upon them, that in a little more they might reasonably expect it would eat up the whole town'.[8] A few years earlier, John Warburton had depicted Brighton with 'the sea having washed away the half of it; whole streets being now deserted, and the beach almost covered with walls of houses being almost entire, the lime or cement being strong enough, when thrown down, to resist the violence of the waves'[9]; and 15 years later, another commentator, writing in 1735, saw Brighton as 'the Ruins of a large Fishing Town'.[10]

This common version of Brighton's pre-resort maritime history[11] has been challenged and refined by Berry. She argues that 'the period when Brighton's economy was dependent on fishing was over before 1680'; that the industry had declined because of changing economics, competition from elsewhere and attacks by foreign privateers as well as the continual and longer term erosion of the shoreline (with some Brighton boats forced to retreat to the estuary of the River Adur a few kilometres to the west as a haven); and that in any event Brighton at this time is better characterised as a seafaring and maritime trading town rather than an important fishery.[12]

By the middle of the fourth decade of the eighteenth century, the first recorded holidaymakers also spent time in the town. Apart from the newly discovered natural delights of the sea and coast, a town like Brighton, that for the wealthy and leisured was reasonably accessible from London, provided buildings to be used for holidaymaking purposes, and local people were available to service the new industry. Dr Russell's arrival, in this perspective, was the catalyst needed to turn the embryonic resort into a rude and healthy infant.

As Brodie and Winter argue in this volume, the Brighton story can be repeated in a number of other coastal localities where sea bathing and the emergent resort industry rescued decaying and increasingly moribund coastal towns. Although there may have been regrets over the loss of custom and tradition, the influx of visitors both at the time and nowadays is typically seen as a progressive and transformative force.

Although the precise timings and processes at work are argued about, it is clear that by the arrival of the first holidaymakers, Brighton beach itself had become a narrow and marginal place fronting onto low cliffs, which were increasingly subject to erosion and consequently threatened the southern limits of the higher town. The surviving fishing industry became more reliant on the use of the Steine, a low marshy valley forming the eastern boundary of the higher town and running down to meet and merge with the beach. Although smaller and less vibrant than a century before, the town's fishing industry was still significant. One observer in 1761 counted 11 large vessels and 57 small fishing boats, employing an estimated 300 fishermen (and, of course, supporting a much larger number of family members). At a time when the town's resident population is estimated to have been between 2000 and 3000 people, the fishing community remained important numerically (they also outnumbered the estimated number of visitors), in terms of their use of space, and economically.[13]

A closer historical analysis presents the old purposes and communities of the town – particularly those centred on fishing – as coming into conflict with newcomers and new activities and, inevitably, losing battles, position, influence and role. Most often, this stance is illustrated through the history of the Steine and the heated disputes between newcomers, seaside authorities and the fishing community.[14] Writing in 1818, one commentator described how the Steine, by then 'the fashionable promenade', was half a century before

> called Stein Field, and nothing more than common wasteland, indiscriminately used by the inhabitants for the repository of heavy goods, sale of coals, boat building, net making, &c. The Steyne was levelled and enclosed, and, as the company invariably promenaded in this field, the nuisances gradually disappeared.[15]

The open space was being transformed from a working area used by the fishing community to a respectable resort promenade ground. Despite opposition from the fishing community, the initial enclosing of the Steine occurred in 1776. But despite these early 'improvements', visitors complained that 'the company, while walking, are frequently tripped up by entangling their feet, and if any one of the barbarians to whom the nets belonged should be standing by, you are sure to be reprobated and insulted for what you cannot avoid'.[16] Such sentiments illustrate the long-term marginalisation of the fishing community by many visitors, officials and commentators alike, serving to represent Brighton fishing people and families as some combination of the natural, simple, poverty-stricken, idiosyncratic, uncouth or roguish. More rarely and sympathetically, a traditional, heroic and proud fishing community is portrayed.[17]

Subsequently, the Steine as a pleasure ground was remade in more elaborate versions through the addition of iron fencing, floral beds and shrubs, paths, lighting, the encasement of a stream, varied commemorative statues and memorials. The promenading space was also increasingly fringed by key resort facilities – for example, baths and circulating libraries – and accommodation was used by the resort's most notable visitors including the Prince Regent's marine villa, a structure remade from a former farmhouse. The further enclosure of the Steine in the 1820s was intensely opposed by the fishing community since it meant the loss of essential space for fishing industry purposes including net drying and boat sheltering (see Figure 5.2).

Using the Steine as their Brighton illustration and evidence from Margate and Hastings, Brodie and Winter, advocates of seeing the early history of resorts as being in part a power struggle over the use of space, argue that in the Brighton context 'in the competition to claim space it was the fishermen who lost out'.[18] Durr seemingly agrees, describing how with the Steine enclosed the only physical presence the fishermen had in the town above the beach was a single capstan on a cliff top. Wanting the space as part of a road-building proposal, the town commissioners decided to remove the capstan. Despite repeated protests from the fishing community, the commissioners persisted with the plan, and in September 1827, the fishermen rioted, with the protest being quelled, in part through the use of almost 100 special constables. As Durr says, 'The fishing industry was thus removed from the town'.[19]

Figure 5.2 The opening of the Victoria Fountain, the *Illustrated London News*, May 1846. The event marked the final stage in the appropriation of the area once used for the Brighton fishing industry and its transformation into pleasure grounds.

Throughout the nineteenth century similar events occurred in other coastal resorts.[20] The story of the appropriation of the Steine by seaside authorities and its transformation into a respectable and regulated space for visitors, and similar events elsewhere in seaside England, while attractive to present-day radical historians, in reality mostly seems to have been unsuccessful rearguard actions against the 'progressive' forces of the holiday industry. Of course, the Steine did not survive as a select promenade. Today, it is a popular seaside park encircled by a traffic-laden road system.

In Brighton (and elsewhere), the evidence presented in the view from Dr Russell's wall plaque seemingly confirms that the rise and growing dominance of Brighton's resort industries in the eighteenth and nineteenth centuries led to the increasing marginalisation and subjugation of the fishing industry and community. But another contemporary Brighton view suggests more complicated and beneficial relationships between the two.

View Two: Brighton Fishing Museum

Leaving Russell's wall plaque and crossing the seafront road to the lower promenade reveals an alternate history of seaside Brighton. There, prominent amongst the clubs, bars, shops and stalls of a successfully regenerated seafront is the Brighton Fishing Quarter (see Figure 5.3). A collection of wooden fishing boats is surrounded by a variety of businesses, some fishing-based, including a fish smoker's and fish and shellfish stalls, and others exhibiting and selling

Figure 5.3 Brighton Fishing Museum and Fishing Quarter, July 2010.

visual art. The centrepiece of the quarter is the Brighton Fishing Museum housed in two sea-front arches. The museum displays, explains and celebrates the complicated history and powerful heritage of the once-important beach-based fishing industry.

The contents of the museum immediately provide intriguing indications of a complex and enduring accommodation between the resort and fishing industries. The conflicts over the use of the Steine and the cliff top capstan are recounted with the added insight that the town authorities provided new facilities and new capstans on the beach. Other material and artefacts for the subsequent decades to the mid-twentieth century suggest that, on the beach and promenades, fishing people and families had a symbiotic relationship with the seaside holidaymakers.

In Brighton, along with a number of other towns on the south coast of England without natural or purpose-built harbours, the beach was a critically important space, serving a dual role as both port and resort. Fishing, commercial and passenger vessels and varied associated activities used the beach and shore as major working spaces while resort visitors increasingly turned to these same sites for health and pleasure. Until the relatively recent past, it was never possible in Brighton for commercial fishing and holidaymaking to use separate spaces.

An accommodation between fishing and holidaymaking is hinted at in the activities of the pioneer visitors to Brighton in search of health, leisure and pleasure. Two of the first adventurers were the Reverend William Clarke and his wife. In the summer of 1736, they stayed for a month in the town, and on 22 July, Rev. Clarke wrote to his friend, a Mr Bowyer:

> We are now sunning ourselves upon the beach at Brighthelmstone ... The place is really pleasant; I have seen nothing in its way that outdoes it ... My morning business is bathing in the sea, and then buying fish; the evening is riding out for air, viewing the remains of old Saxon camps, and counting the ships in the road, and the boats that are trawling.[21]

This is one of the first accounts of seaside leisure and pleasure, as also noted by Brodie in this volume. Rev. Clarke and his wife enjoyed using the beach and sea as well as some of the products of the fishery including buying fish and the visual delights of looking at fishing vessels at work on the sea. Lodging in an older pre-resort Brighton accommodation, almost certainly owned by someone connected to the fishery, Clarke wrote: 'I assure you we live here almost under ground ... But as the lodgings are low they are cheap; we have two parlours, two bed chambers, pantry &c. for 5s. per week.'[22]

This initial accommodation between locals and newcomers quickly developed with the invention of new apparatus for the beach – including bathing machines, beach chairs and benches and pleasure boats – designed to structure and facilitate the seaside holiday for health and leisure and enabled by a pecuniary transaction. The associated businesses seem to have been in (perhaps large) part both controlled and operated by the fishing community. Fishing families owned and ran many of the critically important bathing machines. In addition, the bathing machine guides, an important part of the early bathing ritual helping control how the sea was consumed, were almost all drawn from the local fishing community: Brighton's famed female dipper was Martha Gunn, and the most well-known male bather was John 'Smoaker' Miles.[23]

In the era before municipally owned deck chairs, fishing families supplied and gained an income from the chairs and benches for people to sit on the beach, and they sailed, crewed and provided the pleasure boats. The available evidence suggests that, on many occasions, vessels would be used for both fishing and pleasure boating; there was a seasonal element here with the detritus of fishing being removed and boats washed out at the start of the pleasure boating season.[24]

For the fishing community, the new seaside industries and the emerging resort provided additional opportunities for business and a range of other beach-based and maritime-related occupations. There was a growing cargo (including, e.g. general cargo, food, building materials and coal) and passenger trade. Although passengers alighting from larger vessels (the cross-Channel route was particularly important) were generally rowed and then carried ashore, cargo vessels were typically beached at high tide, unloaded as the tide receded and re-floated on a subsequent high tide. Fishing, though, remained an important under-pinning activity with fish landed on the beach and sold from a beach fish market for local and regional, including London, consumption.

Artists and topographers from the late eighteenth century and photographers from the mid-nineteenth century illustrate how the resort and fishing industries were intricately related throughout the period to the mid-twentieth century (see Figure 5.4–5.7).[25] Early on, typical views of the beach showed fishing boats jostling with bathing machines and holidaymakers gazing at fishermen and the artefacts of the fishing industry. Although the home of the fishing industry – with its array of structures, some made out of old boats – might provide disagreeable sensory experiences for many visitors, for others, it assailed their romantic, picturesque and noble sensibilities in positive ways. Writing towards the end of the nineteenth century, Richard Jefferies, in his distinctive and atypical voice, believed

> The fishing-boats and the fishing, the nets, and all the fishing work are a great ornament to Brighton. They are real … They speak to thoughts lurking in the mind; they float between life and death as with a billow on either hand; their anchors go down to the roots of

Figure 5.4 'The Beach at Brighton – From a Drawing by Miss Runciman, the *Illustrated London News*, October 1859. The town's fishing community provided much of the paraphernalia of the autumn pleasure beach.

existence. This is real work, real labour of man, to draw forth food from the deep as a plough draws it from the earth. It is in utter contrast to the artificial work ... of the town.[26]

Brighton beach was a fluid rather than static entity. It was both a pleasure beach and, despite Shields' assertion to the contrary (see above), a working beach. Its form also continually evolved. The first ineffective coastal defences – wooden groynes – were constructed from the 1720s and subsequently necessarily strengthened and improved. The developing resort industry provided the incentive and financial means for increasingly effective sea defences that, in turn, stabilised the beach and, with an accumulation of shingle, increased its height and depth. In the 1830s, the first concrete sea wall was constructed, and in 1867, the first concrete groyne was built.[27] These large structures also functioned as promenading spaces. Some groynes had additional purposes: one carried lights designed to guide fishing boats safely ashore at night, while some also doubled as sewer outfalls, leading to the surprising if unappreciated prospect of resort visitors sometimes promenading over what they had bodily evacuated in their boarding houses a short while before.

Figure 5.5 The mixed economy of the mid-nineteenth-century Brighton beach, with inshore fishing boats at anchor offshore.

Massive capital spending on new technologies and engineering created further change to Brighton's seafront geography. Away from the older seafront, new elite promenades and coastal gardens and parks provided controlled and segregated exterior places by the sea. The new seaside piers also challenged the status quo. In Brighton, the first of these transitional structures, part landing stage and part select promenade, was the Chain Pier, opened in 1823. The Chain Pier's enabling Act of Parliament emphasised the landing and shipping of people and goods function of the pier but also recognised the 'Purpose of walking for Exercise [and] Pleasure'.[28] The coming of the railways to the town in the early 1840s quickly swept away most of the beach-based cargo trade, and, in any case, there was significant investment in major new facilities in Shoreham and Newhaven harbours, to the west and east of Brighton, respectively.[29]

There were continuing changes, too, in the central section of Brighton's seafront. An inadequate cliff top route, forming the southern boundary to the higher town, was replaced in the 1860s with a new artificial seafront further south, with additional 'improvements' again in the 1880s. This work, by the local authority, also provided at beach level a purpose-built fishing quarter housed in a series of arches supporting the extension of the seafront above. What was to become the permanent

Figure 5.6 Bathing machines and fishing boats side by side and just to the west of the beach-based fish market, *c.* 1880.

home for the industry for the next century served the still large and important beach-based fleet. It included space for the 1813 Fisherman's Society, a school-room and reading-room, a lecture and meeting hall, interior storage and working spaces – at a nominal rent – for fishermen and a purpose-built fish market – the first municipal one on the south coast.[30] The town authorities, albeit with an element of paternalism, were actively encouraging and supporting Brighton's fishery.

The view of and from the Brighton Fishing Museum is then at odds with that from Dr Russell's plaque. The museum – housed in the mid-nineteenth-century lecture and meeting hall – tells the story of a complex and mutually beneficial interrelationship between fishery and resort continuing over many decades. But what of that relationship today?

View Three: Brighton Beach between the Two Piers

A walk from the Fishing Museum across a few hundred metres of shingle to stand on the beach between Brighton's two piers – one a stark burnt out skeleton and the other a throbbing seaside pleasure machine – seemingly reveals yet another perspective. On the beach itself, a few relics of fishing survive, including

Figure 5.7 Brighton beach and the West Pier, *c.* 1900.

rusty winches that once heaved vessels from the sea to the safety of the beach. But the centuries-old tradition of a beach-based fishing industry has disappeared, and there is no visual evidence at all of an active modern fishing industry. On a warm and sunny day, the view is of a beach and sea crowded with people at leisure, but the fishery appears vanquished from Brighton (see Figure 5.8).

To understand what has happened, we need to consider the history of the fishery over the last century and a half in its own terms as well as in relation to resort Brighton.[31] The nineteenth-century Brighton fishing industry was not isolated and separate but instead was part of an economic and social fishery network that sometimes extended beyond the locality and region to the national and even international. The network involved both cooperation and competition between fisheries and fishing ports. Many Brighton boats, for example, were made in Shoreham and Newhaven, both with strong shipbuilding traditions, although others came from as far afield as the West Country and Scotland. The West Country was a favoured source for nets. Technical innovation proceeded apace in vessel construction and the means of power and fishing and refrigeration techniques. The smaller, more regional fisheries were increasingly challenged by the emergence of large national deep-sea fisheries based in ports such as Hull. In turn, the market for fish was radically transformed, particularly with the creation of new consumer preferences and the fish and chip industry being dependent on the

Figure 5.8 Enjoying the sun and beach, Brighton, July 2010.

deep-sea fisheries and refrigeration. International political change, increasingly apparent issues of the depletion of fish stocks, and the need for conservation and changing consumer tastes were subsequent ingredients in the complex mix.[32]

By the mid-nineteenth century, the fishing industry was a relatively insignificant part of the Brighton economy. However, the town still had the largest fleet and the largest resident population of fishermen in Sussex and, in these terms, was surprisingly unchanged from the situation a century before. A mid-1860s Royal Commission into sea fisheries had little to say about the Brighton industry, noting that 'From Brighton the trawlers are of a small class, and fish mainly 10 to 20 miles from the coast, but occasionally within two or three miles'.[33] Nevertheless, there was a considerable local and regional market for fish. In 1892, of the 1192 tons of fish landed on Brighton beach, 39% was consumed locally, with the remainder 'bought inland' by rail.[34] As with other fishing ports, there were sometimes significant annual variations in the size and value of the Brighton catch. Nonetheless, and despite being based on a beach that was also used by the resort, in the last decade of the nineteenth century, the Brighton fishing industry was large in south coast terms. For example, by 1898, Brighton ranked seventh in terms of quantity of fish landed (1324 tons of wet fish) and sixth in terms of the value of its catch (including a little shellfish, a total of £27,657) out of 67 south coast fishing ports or districts listed. It was the fourth most valuable fishing port in Kent and Sussex after Whitstable, Ramsgate and

Folkestone.[35] The value of the fish landed in nearby Newhaven and Shoreham was £3,375 and £8,173, respectively, with – unusually for the majority of south coast ports – almost 50% of the Shoreham value coming from shellfish.

During the twentieth century, the concept of a nucleated and clearly defined Brighton fishery became increasingly redundant. Instead, the Brighton industry developed an increasingly complicated relationship with two nearby fishing stations, Newhaven and Shoreham.[36] In part, this was facilitated by the improved road network and motor transport; in a manner that was impossible even two generations previously, fishermen could live in Brighton and sail out of Newhaven or Shoreham. Newhaven, 13 kilometres by sea from Brighton, and Shoreham, just 10 kilometres away, are both river-mouth ports, mostly useable whatever the weather and able to accommodate far larger vessels than those making up Brighton's beach-based fishing fleet. From insignificant fish landings at the start of the twentieth century, the Newhaven catch grew quickly. By 1913, 1746 tons of wet fish were landed in Newhaven – four times the 443 tons landed in Brighton and many times the 5 tons landed at Shoreham. Part of the explanation may relate to the changing technology of fishing, with Newhaven being able to accommodate larger and far more efficient steam-powered fishing vessels and Brighton still dependent on sail. However, Newhaven fulfilled other roles; its boat-building industry, for example, supplied many of the craft used in Brighton until the 1970s. The fishing industry in all three places was functionally intertwined, with some fishing vessels, for instance, using Brighton beach and, as occasion arose, either Newhaven or Shoreham harbour.

The Brighton catch exceeded that of both Newhaven and Shoreham during the 1920s. However, from the early 1930s through to the 1960s, while the value and size of the Brighton catch fell, that of Shoreham remained insignificant but that of Newhaven steadily increased. In 1934, 194 tons of wet fish were landed on Brighton beach and 208 tons at Newhaven; the respective values of the catch, including shellfish, were £8,036 and £10,576. By then, Newhaven had been officially elevated to stand alongside Brighton as a 'major' fishing station, a designation that Brighton was to relinquish a few years later in the decade. In 1938, 254 tons of wet fish were landed at Newhaven and just 130 tons were landed on Brighton beach (and 8 tons at Shoreham).

There were other significant changes impacting on the Brighton fishing community's livelihood during the interwar years. The traditional beach-based resort occupations available to the fishing community were eroded by the coming of the sun to seaside holidaymaking, the modernisation of resorts and changes in fashion and regulation. Increasingly, the bathing machine business and all it had involved disappeared as holidaymakers found (and municipal authorities approved of) new ways of using the beach and the sea. The local authority controlled and policed the beach, supplying and charging for the ubiquitous deckchairs that had replaced the miscellany of chairs and benches previously supplied privately. One technical innovation – the introduction of petrol-powered inboard engines in the 1920s and 1930s – at least liberated the Brighton fleet, both in fishing and Skylark pleasure boating mode, from the tyranny of wind and sail alone.

The Second World War hurried the movement of the fishery from the beach. The seafront was made into a defensive wall against possible invasion. Most Brighton boats were moved to Newhaven or Shoreham from where they could still fish, albeit under extreme restrictions. Those that remained rotted away.[37]

By 1949, just 30 tons of wet fish were landed on Brighton beach, valued (with shellfish) at £4,677; for Newhaven, the figures were 300 tons and £23,011 and for Shoreham, 12 tons and £1,280. Early in the post-war period, there were more lucrative returns to be had by Brighton fishermen from pleasure trips. By 1963, boats landing at Brighton (on this occasion including the insignificant figure for Shoreham) caught 45 tons of fish; those at Newhaven, 203 tons; the respective values were £8,770 and £37,890. As if to signify the demise of the Brighton industry, the fish market on the beach 'hard' between the two piers was closed by the Council in 1960 on the grounds of poor hygiene and moved slightly inland and indoors. The new market did not survive.

However, there were two other significant post-war developments. First, with its eastern boundary including part of the Brighton Council area, the Shoreham Port Authority increasingly welcomed fishing vessels, both small day boats and larger deep-sea vessels, to the harbour (see Figure 5.9). Although for the larger vessels seeking the calm water, security and facilities of a significant port there was a lock to negotiate, the port had much to offer, and by the end of the twentieth century, it had established itself to the degree that it was officially classified as a 'major' national fishing port – the only Sussex port categorised in this way since Newhaven lost the designation a few years earlier. Second, from the mid-1960s, plans were developed for the construction on the eastern built-up boundary of Brighton of what was claimed to be the largest marina in Europe. Brighton Marina (see Figure 5.10), part designed for pleasure boats and part for property development, opened in 1978. Apart from pleasure boat moorings, it quickly became home to a number of day-fishing boats. The marina also offered a reinvented version of the earlier pleasure boating from Brighton beach including sailing lessons, angling and sightseeing trips.

Figure 5.9 Part of the deep-sea fishing fleet based at Shoreham Harbour, although still within the Brighton and Hove local government boundary, May 2008.

Figure 5.10 Brighton's inshore fishing boats, sandwiched between holiday apartments and pleasure boats, Brighton Marina, May 2008.

A Glance Back

Looking back over the promenade and the three views reveals contradictory scenes and perspectives about the relationships between port – and particularly the fishery – and resort in Brighton from the 1730s to the present day. However, the most useful approach to this promenade has been to look behind the surface appearances. This suggests that a continuing and mutually beneficial accommodation and enduring resilience characterise both resort and fishery.

In 2008, 138 tons of fish and shellfish valued at £453,267 were landed in Brighton, 1027 tons valued at £1,895,920 at Newhaven and 2931 tons valued at £4,791,271 at Shoreham.[38] As argued above, the three ports are best seen as a single integrated fishing station. A year later in 2009, of the 24 officially listed English and Welsh fishing ports, Shoreham – landing 5088 tons of fish and shellfish worth £7.1m – was fourth largest in terms of the value of the catch, exceeded only by Plymouth, Newlyn and Brixham.[39] Shellfish and particularly scallops – recently emerging as fashionable seafood in demand throughout the Western world – dominated the Shoreham catch, in terms of weight (88% of the total) and value (76% of the total). The single most important present-day seafood export from Brighton is of whelks to South Korea.[40] In March 2010, 58 fishing vessels were based in Brighton or Shoreham. Twenty-three fishing boats of 10 metres length or under used Brighton – mostly the marina – as their home port and another 30 boats of the same length and 5 over 10 metres had Shoreham as home port; other large vessels, including some from Scottish ports, spent much of the year based at Shoreham.[41]

In large part, through the preference and choice of the fishing community, facilitated by a range of social, economic and technical changes, the centuries-old Brighton fishery gradually vanished from the beach over the last century. However, the disappearance is better seen as a transformation into a port-based fleet located at the marina, Shoreham and Newhaven. The marina – a major project designed to help regenerate Brighton as a resort – freed fishermen with their boats in Brighton from their dependency on the beach. At last, they had a secure and safe saltwater port. The reinvention of resort Brighton helped save its fishing industry.

The transformation of the fishery made the mid-nineteenth-century Fishing Quarter increasingly redundant. These events coincided with the decline, for two decades from the 1960s, of Brighton as a resort. The indicators included falling visitor numbers, the closure of seaside attractions and the increasing dereliction of much of the holidaymaking built environment.[42]

Through human agency and as part of a large-scale scheme to regenerate a decrepit central seafront, in the 1990s, the Brighton Fishing Museum and a reinvented Fishing Quarter were established as one of the 'string of pearls' of new seafront attractions.[43] Making, crafting, displaying and selling – the variety of Fishing Quarter enterprises accommodated both members of Brighton's fishing community and newcomers lured to the beach by location or the anticipation of profit. The museum preserved, interpreted and explained the past. It re-invented forgotten ceremonies – such as the blessing of the nets (see Figure 5.11) – for touristic purposes. In this sense, the fishing industry had been transformed into heritage. Most recently, the museum has included an exhibition on the now derelict West Pier; the artefacts of port and resort in the past are displayed side by side. It also provided new commercial opportunities including the sale of local fish and shellfish landed at the marina. As the centrepiece of a successful new tourist attraction, the remains and memories of the old beach-based fishery helped to save and renew the seafront and transform it into a fashionable twenty-first-century seaside promenade and leisure space.

Figure 5.11 Celebrating Brighton's fishing industry heritage. Spectators watch the Blessing of the Nets ceremony, May 2010.

Notes and References

1. Special thanks are due to Andy Durr for sharing his knowledge of the Brighton fishing industry and Fishing Museum. Holly Gray provided valuable ideas and photographs and both Sue Berry and Kathryn Ferry provided essential information and intriguing leads. Some of the material in this essay is developed from Fred Gray, *Designing the seaside: architecture, society and nature* (London, 2006).

2. Corbin, Alain, *The lure of the sea: The discovery of the seaside in the western world 1750–1840* (Cambridge, 1994)

3. Still grander claims have been made of Russell as the inventor of the Western seaside. For a discussion see Gray, *Designing the seaside.*

4. Ruth Manning-Sanders, *Seaside England* (London, 1951), p. 16.

5. Fred Inglis, *The delicious history of the holiday* (London, 2000), p. 41.

6. Rob Shields, *Places on the margin. Alternative geographies of modernity* (London, 1991), p. 81.

7. The early development of Brighton as a resort is best described in: Sue Berry, *Georgian Brighton* (Chichester 2005), an expanded and developed version of Sue Farrant, *Georgian Brighton 1740–1820* (Brighton, 1980). See also John and Sue Farrant, *Brighton before Dr. Russell: An interim report* (Brighton, 1976).

8. Quoted in Eric Underwood, *Brighton* (London, 1978), pp. 53–4.

9. Quoted in Ian C. Hannah, *The Sussex coast* (London, 1912), p. 184.

10. Quoted in Farrant, *Brighton before Dr. Russell*, p. 3.

11. See, for example, the influential Edmund W. Gilbert, *Brighton. Old ocean's bauble* (London, 1954) pp. 50–3.

12. Berry, *Georgian Brighton.*

13. This section draws on Berry, *Georgian Brighton*, Underwood, *Brighton* , Gilbert, *Brighton* and Andy Durr, 'Introduction', in Andy Durr, ed. *Charles Fleet's, The Brighton fishery, in the mid nineteenth century* (Brighton, 1994), pp. i–vii.

14. This section draws on Durr, 'Introduction', Gray, *Designing the seaside* and Gilbert *Brighton.*

15. Charles Wright, *The Brighton ambulator* (London, 1818), pp. 47–8.

16. Underwood, *Brighton*, p. 76.

17. Both portrayals are detailed in Durr, 'Introduction'.

18. Allan Brodie and Gary Winter, *England's seaside resorts* (Swindon, 2007). p. 50. This fits into wider patterns of conflict in eighteenth- and early nineteenth-century England: E. P. Thompson, *Customs in Common* (London, 1991); A. Randall, *Riotous Assemblies: Popular Protest in Hanoverian England* (Oxford, 2006).

19. Durr, 'Introduction', p. iii.

20. Steve Peak, *Fishermen of Hastings. 200 Years of the Hastings fishing community* (St. Leonards on Sea, 1985), p. 32, and John Field, 'The Battle of Southsea', *The Portsmouth Papers*, 34 (1981).

21. Quoted Wright, *Brighton Ambulator*, pp. 24–5.

22. Ibid, p, 25.

23. Kathryn Ferry *Sheds on the Shore. A tour Through beach Hut history* (Brighton, 2009), p. 73.

24. QueenSpark Books, *Catching stories. Voices from the Brighton fishing community* (Brighton, 1996).

25. David Beevers, ed, *Brighton revealed through artists' eyes c. 1760–c.1960* (Brighton, 1995) and Philippe Garner, *A seaside album: photographs and memory* (London, 2003).

26. Richard Jefferies, *The pageant of summer* (London, 1979), p. 111.

27. Timothy Carden, *The encyclopaedia of Brighton* (Lewes, 1990)

28. John George Bishop, *The Brighton Chain Pier: In memoriam* (Brighton, 1897), p. xviii.

29. J.H. Farrant, *The harbours of Sussex, 1700–1914* (Brighton, 1976) and H.C. Brookfield, 'Three Sussex ports, 1850–1950', *Journal of Transport History*, II, 1955.

30. Durr, 'Introduction'.

31. Much of this section is based on the oral testimony provided in QueenSpark Books, *Catching stories*.

32. See, for example, Ruth H Thurstan, Simon Brockington, Callum M Roberts, The effects of 118 years of industrial fishing on UK bottom trawl fisheries', *Nature Communications*, 1, May (2010), pp. 1–6; John K Walton, *Fish and chips and the British working class 1870–1940* (Leicester, 1992) and Robb Robinson, 'The Development of the British Distant-Water Trawling Industry, 1880–1939', in David J Starkey and Alan G Jamieson, eds., *Exploiting the sea* (Exeter, 1998).

33. HMSO, *Report of the Commissioners appointed to inquire into the sea fisheries of the United Kingdom* (London, 1868), p. 65.

34. HMSO, *Statistical tables and memorandum relating to the sea fisheries of the United Kingdom in the year 1892* (London, 1893).

35. HMSO, *Statistical tables and memorandum relating to the sea fisheries of the United Kingdom in the year 1898* (London, 1899).

36. This figures in this section are based on the annual sea fisheries returns for the United Kingdom.

37. QueenSpark Books, *Brighton behind the front* (Brighton, 1990).

38. Personal communication, 1 April 2010, Sussex Sea Fisheries Committee.

39. Fisheries Statistical Unit, Marine and Fisheries Agency, *Monthly return of sea fisheries statistics for England, Wales, Scotland and Northern Ireland. Provisional. Month of December 2009* (London).

40. Personal communication, 30 March 2010, Neil Messenger, proprietor, Sea Haze fishmongers, Brighton.

41. The figures are calculated from: Department for Environment, Food and Rural Affairs (DEFRA) Fisheries Statistics Unit, The United Kingdom Fishing Vessel List (excluding islands) as at 1 March 2010, contains registered and licensed vessels of 10metres and over overall length and DEFRA Fisheries Statistics Unit, the United Kingdom Fishing Vessel List (excluding islands) as at 1 March 2010, Contains registered and licensed vessels of 10metres and under overall length. http://www.mfa.gov.uk/statistics/vessellists.htm, Accessed 22 March 2010.

42. Gray, *Designing the seaside*, p. 82

43. Andy Durr, 'The making of a fishing museum', *History Workshop Journal*, 40 (1995) pp. 229–38.

Chapter 6

From Port to Resort: Tenby and Narratives of Transition, 1760–1914[1]

PETER BORSAY

The rise of the seaside resort would seem to embody many of the processes that are associated with that period of change in the later eighteenth and nineteenth centuries that has come to be characterised as the Industrial Revolution. Resorts were new and transformative. They had no real precedent (other than perhaps the inland spa). They had the capacity to convert a virgin site or established coastal settlement into something entirely different in character and appearance. Moreover, they appeared to epitomise the essence of the Industrial Revolution: *specialisation* and *differentiation* of production and location. Read in these terms, resorts became a highly specialised type of urban settlement focused on the provision of health and leisure and were part of a sharply differentiated but integrated urban network.

How did the transformation suggested here work out in practice? How did an individual location, such as an established port, negotiate the potentially dramatic changes it faced? Was it simply a matter of swapping one *modus operandi* and identity for another? To address these questions, the following essay takes as its case study the Pembrokeshire town of Tenby (see Figure 6.1), located in south-west Wales, on the north-western edge of the Bristol Channel. By the time of its birth as a resort, it was already a well-established port with a long history. Founded after the Norman Conquest, if not before, it was one of the more substantial and important urban settlements of medieval and early modern Wales, with overseas trading connections (in the fourteenth century, it imported more wine from Bordeaux than any other Welsh port), an independent civic community incorporated from 1581 and a mature built environment encased in impressive fortifications.[2] Tenby came, in other words, with a history. It was not a new town, being constructed, like some resorts, on a virgin, or near virgin site. There was a genuine transition to be negotiated between port and resort, between past and future. There are narratives, both of change and continuity, to be recovered.

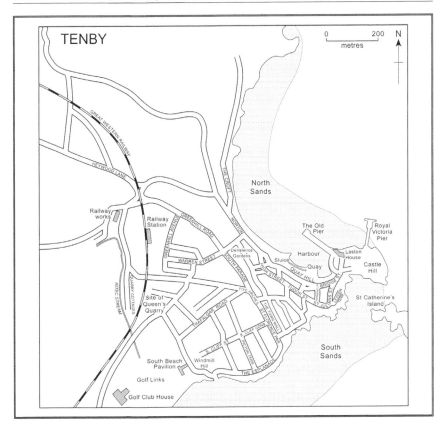

Figure 6.1 Map of Tenby *c.* 1940.

The Rise of the Resort

Tenby emerges as a resort in the late eighteenth century. The monument erected in St Mary's Church to the 82-year-old 'PEGGY DAVIES. Bathing woman 42 years to the Ladies who visited TENBY', following her death while at her post in 1809, would suggest that the sea waters were already accommodating fashionable clientele by 1767. However, a visitor to the town in September of the same year fails to make any mention of the bathing, though that may be due to the lateness in the season.[3] Another visitor in late August 1787 could write, 'Mrs M. unfortunately bruising her hand the idea of bathing any longer is given up', and the comment in James Baker's *Picturesque Guide* of 1791, 'there is latterly some resort here [Tenby] by strangers for sea bathing, for which the sands are particularly commodious, and there are other conveniences provided them', suggests that the town had already acquired a reputation as a resort and was adapting to meet the needs of bathers.[4] This would support the notion of Tenby being part of the second wave of resort foundations, as the new fashion spread from its birthplace, primarily the south-east of England (though Scarborough hosted sea bathing by at least the 1730s) in the early eighteenth century to the more 'peripheral' regions of Britain, during the latter part of the century.[5] In Wales, Tenby was joined at this time by the two further pioneer resorts of Aberystwyth – where there is evidence of the use of the sea waters from 1767[6] – and Swansea.[7]

The reasons for Tenby's rise as a resort are twofold. On the negative side, by the late eighteenth century, the town had been trapped in a prolonged economic depression, for which the origins date back to at least the seventeenth century, with the decline of its indigenous fishing industry, a loss of overseas trade, the relative lack of prosperity of its rural hinterland and competition from neighbouring industrial districts and towns. The consequence was a town in a state of human and physical decline.[8] A visitor of 1767 noted that 'within these Walls are the most compleat Ruins of an old Town, the Houses still standing, tho' unroofed ... Certain it is, that there has been at least 4 times the number of Houses in it that there now is', and in 1812, Charles Norris – whose drawings graphically recorded many of the decaying buildings – observed that 'till the last twenty, or twenty five, years the town was almost entirely deserted, excepting by the poorer classes, and a few respectable tradesmen'.[9] Bad as all this was for Tenby, and there may be some exaggeration in this narrative of decay (there is little to suggest that its population was any less in 1801 than in 1670),[10] it meant that here was a town eager to find a new economic role.

On the positive front, Tenby possessed two critical physical attractions which resonated closely with the picturesque and romantic ideal of a resort. Situated dramatically on a rocky promontory, and with a prospect shaped in the near distance by Caldey Island, Giltar Point and Monkstone Point, and in the further distance by the Gower and Worms Head, there were varied and spectacular marine views (see Figures 6.2 and 6.3). The sea was one of the great cultural discoveries of the eighteenth century,[11] and Tenby constituted a perfect platform from which to observe it. The town was also blessed with a suite of fine sandy beaches from which visitors could take the health cure. Contemporaries were quick to commend the natural assets of the town. 'Tenby is very prettily situated on a Cliff by the sea side having at its base a sandy beach', wrote one observer; another found that 'the town of Tenby is but small, displaying very little more imagery than the sea view, which is here enjoyed in very high perfection ... The water is clear & strong, & a gradually declining beach is found hardly to be equalled any where'; another discovered 'the situation of Tenby is fine – upon a rock of considerable elevation, so as to afford delightful sea views, with the coasts of Pembroke and Carmarthen-

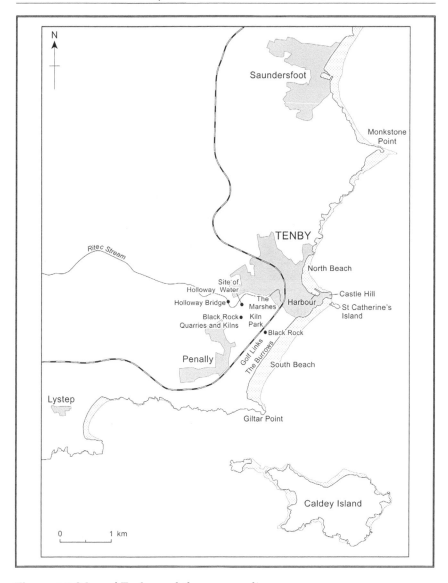

Figure 6.2 Map of Tenby and the surrounding area.

shire in the distance. The sea here is formed into a fine bay & the town extends in a sort of imperfect crescent form on the Cliffs'; and an early guidebook noted, 'This town is very singularly situated, and presents an admixture of woods, rocks and houses that have the most romantic effect. Its views by the sea are particularly striking. The beauty of the situation, and the fine sandy beach, have raised Tenby to a public place of no small

Figure 6.3 Tenby from the North Cliff, W. Golding, oil painting, 1799.

reputation'.[12] To these environmental advantages must be added the fact that Tenby was a working town, with the sort of access to resources – food, accommodation, shops, public services, servants etc. – that an undeveloped site, particularly before the arrival of the railways, would not possess or could not easily acquire.

For much of the nineteenth century, Tenby's development as a resort was steady but not spectacular, though with some slackening in the years before the First World War. The census return for 1801 records a population (for the In-Liberty) of 844; by 1831, this had increased to 1942 people, by 1851 to 2982 people, and by 1881 to 4750 people, dropping a little to 4368 people by 1911.[13] It thus remained a small town throughout the period. In 1851, it ranked number 36 by size in the league table of English and Welsh seaside resorts; by 1911, it had fallen to number 74 as more dynamic resorts – such as Rhyl in north Wales, catering to a broadening market – overtook it.[14] The first substantial phase of investment in the resort occurred at the turn of the nineteenth century. Already in 1781, the council had agreed to lease St Julian's Chapel on the pier to John Jones, 'Bachelor of Physic' of Haverfordwest, 'for the sole purpose of Constructing Baths and other Contrivances'. This venture was far eclipsed by the impressive establishment constructed by Sir William Paxton, a Carmarthenshire banker and landowner, on property beside the harbour, which he had leased in 1805 from the council for the purpose of building a 'bathing house' (see Figure 6.4).[15] This was opened in 1810 and 'fitted up in a superior style'. In addition to

Figure 6.4 The Bath House, Tenby, opened 1810.

the sea baths, there was 'a handsome coffee room', which appears to have doubled as the town assembly room and 'a spacious vestibule for servants and attendants to wait in'. In addition to this flagship building, managed by a couple imported from Bath, Paxton also appears to have built a hotel (the *Albion*) alongside the baths, together with livery stables and coach houses, reconstructed a house on St Julian's Street as the Globe Inn (1807), remodelled the road linking the harbour to the town and overcome one of the major inconveniences faced by visitors in summer – a shortage of water – by building at 'sole expense … two reservoirs, which are never likely to be deficient of that necessary article'.[16]

This was a substantial and calculated investment, spread across facilities and infrastructure, and provided the necessary boost to kick-start Tenby's serious development as a resort. Already by 1813, *The Cambrian Traveller's Guide* could pronounce that 'in the year 1790 the place [Tenby] was almost deserted, but since that time it has become a favourite resort to the fashionable and luxurious'; two years later, the town was said to have 'been greatly improved in its internal state since it became a place of public resort', and in 1831, one visitor observed that 'vast alterations have been made in the last 30 years, new building in every direction & apparently every convenience as regards provisions &c. for families visiting the place'.[17] During the first half of the nineteenth century, the town acquired the essential package of ingredients for a fashionable resort: good indoor and outdoor bathing facilities, assembly rooms, a theatre, markets and shops stocked with fresh food and luxury products, a circulating library, formal promenades, a network of informal walks and excursions in the neighbouring environs and region, comfortable lodgings and residences and a dedicated guidebook to inform visitors and structure their expectations.[18] The provision of suitable accommodation was part of the wider process by which the built environment was upgraded, a critical factor in Tenby's case given the long-term neglect of its buildings before the nineteenth century. Alongside the construction of high-status terraces – echoing those of London, Bath and Brighton – such as the Croft (from the 1830s) and Lexden Terrace (1843), there was the refurbishment and remodelling of the inherited landscape, including the demolition of buildings and fortified gates that obstructed the flow of traffic, as when the Middle Row was razed to open up Tudor Square in 1836.[19] Change was facilitated by the wide-ranging Tenby Improvement Act of 1838, and the overall environment was enhanced by vital work on the public infrastructure of water supply, drainage, sewage and lighting.[20]

Despite the pace of change, Tenby's development up until the mid-nineteenth century was very largely confined to the space within the medieval walls, the immediate vicinity of the walls, a strip of extra mural growth to the north along the Norton and the Croft and the fashionable small suburb of Heywood Lane.[21] This left a good portion, indeed the majority of the promontory on which the town sat, undeveloped. There is little to suggest that the failure to establish an early railway link to the town curtailed expansion. There were good sea communications with the town, hampered only by tidal and physical restrictions on docking and eventually relieved by the construction of an iron pier in the 1890s. Moreover, it could be argued that the resort's success depended upon maintaining a degree of exclusivity, which may be why in 1846 the council determined to oppose 'in every possible manner' the project to bring the Tenby, Saundersfoot and South Wales Railway 'within the limits of the Parish of Tenby ... as they deem a station at Saundersfoot sufficiently near enough to Tenby for the public convenience'.[22] However, it is no coincidence that the establishment of a rail link to Pembroke in 1863, and then to the main line at Whitland in 1866, coincided with the extensive developments to the south and west of the walled town.[23] In the 1860s, *Allen's Guide to Tenby* referred to

> magnificent sites for building purposes along the South Cliff [that] might be profitably converted into terraces of stately residence if any spirited builder would turn his attention to the requirements of this beautiful spot ... having a sea view of unsurpassed extent and beauty and a solid mass of limestone to quarry foundations from.[24]

Figure 6.5 The Esplanade, Tenby, 2010.

In late 1868, the council had considered the removal of the South Gate 'being an obstruction and hindrance to the extension of the Town and the building of the houses on the South Cliff Estate', and early the following year, it commissioned and adopted a plan to lay out building plots in the Rope Walk Field abutting the future Esplanade (handed over as a public thoroughfare in 1875, see Figure 6.5).[25] Over the next three decades, there emerged what was effectively a new town, since not only the sea-facing south end of the promontory filled in with fashionable houses and hotels but also the area to the west and north of it with less prestigious properties, many of which housed service workers. Already by 1882, F.B. Mason could claim that 'since the opening out of the South Cliff, the Parade', running under the west wall, and bisecting the old and new towns, 'has become [one] of the leading thoroughfares of the town', and five years later, it was pronounced at a public meeting that it was in the direction of the Esplanade that the town should grow.[26]

Physical expansion was accompanied by growing investment in cultural capital. This meant not only public facilities such as the Public Hall in Warren Street, the skating rink opened in Picton Road in May 1877 and converted into the New Victoria Pavilion by 1894 and the De Valence Gardens opened in 1904 (see Figure 6.6), but also several non-conformist churches and a plethora of clubs – with varying degrees of exclusivity (the Tenby and County Club, opened in 1877 on the Croft,

De Valence Gardens, Tenby   I am enjoying myself immensely
                                                                    here

Figure 6.6 The De Valence Gardens, Tenby, opened 1901.

operated a blackballing system)[27] – many of which were associated with sporting activities such as tennis, archery, croquet, cricket and golf.[28] Such investment reflected, in part, the fiercely competitive world of the late Victorian and Edwardian resort in which Tenby was struggling to retain its position. It was also a measure of the need to provide visitors and residents with organisational structures and identities in an environment in which the leisured population of the resort had expanded beyond the level at which a single group – the 'company' – and programme could cater to all their interests.

The proliferation of clubs and institutions may also reflect Tenby's need to negotiate the problem posed in the late nineteenth century by the widening social access to a holiday – whether short or long term – away from home. The resort's success was built upon attracting a high-status clientele. In August 1831, one visitor could observe that the company was 'at this time very thin, perhaps owing to the present sitting of Parliament', and one aged inhabitant reminiscing in the 1930s about the town 80 years earlier described how 'all the big houses ... were occupied by the nobility who had their carriages and horses and servants in livery ... During the winter they gave balls and dinners'.[29] Although growing numbers of the mainstream middle class were finding their way to the resort, the polite and gentrified social tone was one that Tenby hung onto resolutely until the First World War and beyond. In 1881, a guidebook congratulated Tenby on being a resort that affords to the statesman, the literary man or the man of business, a perfect haven of rest and pleasure, and the very fact of its being far away from any large town says a deal in its favour; it is spared the society of the noisy, clamorous 'Arry and his associates whom we meet at Ramsgate, the trains are not arriving at every half-hour bringing shoals of Jemimas and Sarah Ann's bearing their indispensable reticule full of buns, beer and oranges.[30]

In reality, the geographical and social profile of Tenby's market was changing; the powerful links it had with Bath and the fashionable West Country were declining by the late nineteenth century, and day trippers – albeit of the more respectable type – from the dynamic South Wales industrial areas were beginning to make a mark, particularly during periods such as bank holidays and for events such as the Tenby Regatta and Races, when special excursion steamers and trains would be mounted.[31] A sign of these changes was a letter received by the council in 1912 from the Aberavon Shop Assistants' Committee informing it that a party of 850 intended to visit Tenby for its 'annual marine trip' and asking for 'a list of hotels and restaurants that cater for Daily Visitors. Their reason for asking this is that they find that Tenby has a bad reputation in the matter of catering'. The tensions that such visitors created among the more traditional class of visitors can be gauged from another letter of 1898 complaining about the insolent behaviour of the operative organising the chairs at the bandstand and feeling it necessary to make it clear to the council that 'I am no mere "tripper". I may venture that the above estate [indicated in the letter head] is my own property and that I am a later (*sic*) Senior Chaplain in H.M.'s Service'.[32]

The Maritime and Industrial Town

The trajectory of Tenby's development in the nineteenth century would seem clear enough, and encourages a narrative of transition that emphasises the rapid and irreversible replacement of port by resort. Indeed, if some of the guidebooks are taken at face value, the process would appear to have been concluded quickly. In 1843, Mary Bourne could declare the town to possess 'no trade of any kind, but is indebted for its prosperity to the resident gentry of the town and neighbourhood, and the numerous influx of visitors'.[33] However, other sources, and other ways of looking at Tenby's nineteenth-century history, while not challenging the long-term trajectory of its development, suggest a process of transition that is more nuanced and complicated and one that sees elements of continuity between the town's traditional and new economic roles.

Tenby's medieval and early modern prosperity, before decline set in, had been based upon the working (as opposed to recreational) sea, either as a medium for the transport of goods and people or as a source of food through fishing. At the heart of this economy, and its physical interface with the town, was the harbour. During the nineteenth century, this remained an important and vital area of activity and the subject of regular maintenance and improvement. Although in 1798, the quay was called 'tolerably commodious' and, in 1819, a visitor noted that 'a pier has been erected, which will protect vessels in all weathers', Charles Norris in 1832 pronounced the pier and harbour to be 'in a most deplorable state of neglect'.[34] The Tenby Improvement Act of 1838 included a provision for 'regulating and maintaining the Harbour and Pier',[35] and in 1842, the *Carmarthen Journal* reported that

> considerable improvements are in progress in the harbour and pier of this increasingly fashionable town. A new pier is to be constructed on the western side of the entrance to the harbour, and arrangements effected to enable two steam-packets at the same time to lie close to the landing place, and the passengers to step at once to the shore: a proceeding infinitely more agreeable than the delay and confusion often arising from the necessity of being landed in boats, which, hitherto, has been very frequently the subject of complaint.[36]

Regular improvements were undertaken in the second half of the century (the sluice was added in about 1869, and the two Quay Hills were macadamised three or four years later), but the most significant innovation in the period was the construction and extension, to the west of the harbour on Castle Hill, of the iron landing pier, formally opened by the Duchess of York in 1899.[37]

Much of the improvement to the harbour and its vicinity was geared to improving tourist access to Tenby, particularly once steamers were regularly in operation. Although this reflected the rise of the resort, servicing the needs of visitors for access by sea (a demand not undercut in the short term by the arrival of the railways), it also sustained and promoted the viability of the harbour and the town's role as a working port. In some respects, this compensated for Tenby's declining impor-tance as a centre of trade. As early as 1812, Charles Norris was unequivocal about the town's commercial demise:

> In Tenby there is no trade. The vessels which are seen here, with two or three exceptions, belong to other places and proprietors, and frequent the harbour either for shelter, or till they can be laden with coals at Sandisfoot, an exposed and, with some winds, a dangerous bay within two miles.[38]

In fact, the situation was not quite as bleak as Norris would have it. He underestimated the extent to which Tenby was able to exploit its position within a buoyant cross-channel economy. In the early nineteenth century, a number of smacks were plying their trade with Bristol, exporting small volumes of grain.[39] The town also needed to import raw materials and food, if only to service its own growing population of residents and visitors. In the mid-nineteenth century, before the establishment of the rail connection, an inhabitant recalled that

> most of our provisions were brought to the town by the sea. ... The discharge of these vessels ... would commence as soon as the ships were made fast to the pier, when about half a dozen quay men would be ready with their trucks to remove the Tenby goods to the Tenby stores which were situated at the bottom of Pier Hill, under the old Assembly Rooms. ... Several sailing ships used to bring cargoes of coal, potatoes and building materials as well as baulks of timber.[40]

In the late nineteenth century and early twentieth century, coal was still being shipped in from Lydney in Gloucestershire (see Figure 6.7) – locally mined fuel was not suitable for domestic use – and market garden produce was shipped from Caldey Island, and between 1913 and 1937, the Bristol grain merchants, Spillers and Bakers Ltd, were leasing one of the stores by the harbour.[41]

Late Victorian and Edwardian photographs of the harbour and surrounding area show it bristling with the masts of dozens of sailing vessels, in some cases perhaps up to 50 craft (see Figure 6.8).[42] Most of these were fishing boats and are a striking testimony to Tenby's role as a fishing port.[43] In the late eighteenth and early nineteenth centuries, the

Figure 6.7 The 'Rainbow', 8 December 1909, delivering coal from Lydney, Gloucestershire.

Figure 6.8 Tenby Harbour, trawlers, *c.* 1905–1910.

indigenous industry was on the decline with the rich local fishing grounds being trawled by boats from south Devon. In 1810, the *Tenby Guide* noted that 'a fishery was once carried on with considerable spirit, but has of late years been much on the decline', and two years later, Charles Norris recorded how 'ten or twelve stout smacks from Brixholm and Dartmouth, lie here during the whole summer, and carry the produce of their industry to Bristol. Scarcely a single decked vessel is employed in fishing from the town'.[44] Sheltering and servicing the Devon boats brought business to the harbour and town, but it failed to fully exploit a local resource. From the middle of the century, this began to change with the revival of the native fleet[45]; by 1891, the register of fishing boats indicates that at least 14 first-class and 7 second-class trawlers were owned at Tenby.[46] This was the peak of the trawling industry. It soon went into decline. Between 1899 and 1914, the volume of fish landed in the town dropped from 6800 cwt to 1596 cwt. The appearance of steam trawlers in the Bristol Channel from the 1880s and the inability of Tenby to compete with the dock facilities at Cardiff, Swansea and the rising Milford in handling these boats marked the demise of the local industry.[47] However, at its height, it had a considerable impact on the harbour area and town. The census of 1881 shows large numbers of fishermen and their families (see Figure 6.9) packed into the warren of streets and lanes surrounding the harbour, dominating their occupational profile. The Quay had 8 fishermen; Quay Hill, 16; Bridge Street, 10; and Sergeant's Lane, 11.[48] The industry also

Figure 6.9 Fishermen in front of the Albion Hotel, Tenby.

impacted on other trades that clustered around the harbour. In 1898, Thomas George, shipwright and boat builder, complained to the council about an order to clear the Sluice Wharf of all boats and timber:

> I have no place where I can keep my timber for trawl beams &c. except on some part of the wharf ... The trawlers and boats are constantly requiring new trawl beams &c. I am obliged to keep timber in stock and if they cannot get served they will have to get their work done elsewhere which will be an inconvenience to them and a loss to me and the town.[49]

In 1818, Tenby was described in one of its earliest guidebooks as 'having little trade, and being in the neighbourhood of no works or manufactories ... [it] is exempt from the bustle, and the frequent scenes of riot by which towns where much trade is carried on, are rendered ineligible as places of residence'. Similar sentiments were expressed in *Allen's Guide* of 50 years later when it described the town as 'untainted by ... unwholesome smoke from works, and free from the noise and bustle of a commercial seaport'.[50] In portraying Tenby as insulated from the impact of the industrial and commercial forces that were reshaping Britain, the guidebooks were being a little economical with the truth. During the late eighteenth and nineteenth centuries, Pembrokeshire 'experienced a miniature industrial revolution of its own' – if hardly on the scale of its south Wales neighbours – based primarily upon the extractive industries, shipbuilding and some metal working and making use of its extensive coastal and inland waterways.[51] An important element of the county's industrial economy was the exploitation of the coalfield of which the eastern limb bordered directly on Tenby.[52] Although in the eighteenth century Tenby had exported coal and culm directly from its harbour, by the early nineteenth century, this trade had been substantially reduced, and Saundersfoot – Tenby's immediate neighbour to the north (and hidden from the town by the headland of Monkstone Point) – emerged as the principal coal port, with its harbour completed in 1835.[53]

Coal and limestone were the key raw materials in the production of lime, widely used in agriculture and the building industry. With limestone unavailable in west Somerset, north Devon, Cornwall, north Pembrokeshire and West Wales,[54] the rich veins of limestone that lay immediately to the south of Tenby – in combination with the coal to the north – provided the basis for a buoyant cross-channel as well as local trade. Limestone quarries existed on Caldey Island, and at Lydstep, Giltar Point and Black Rock[55] – all within hailing distance of Tenby – and even in the town itself. The Ordnance Survey map of 1888–9 indicates the presence, on the southern side of the rocky promontory on which the town sat, of four quarries (Queen's Quarry, one in the Greenway area, another below the new railway station and another beneath Battery Road) and three limekilns (two above the old railway station and another under Windmill Hill).[56] The limekilns are described as 'old', but the quarries are not specifically designated as 'disused', and it is unclear whether they were being exploited for building materials (a possibility given the property development in the area at the time)[57] or lime. The industrial character of the area was reinforced by the presence of two railway stations (the original terminus had been converted into 'railway works' after the construction of the new higher level station in 1871) with attendant track and yards, a slaughterhouse, timber yards and a disused steam corn mill. The census returns for Chimney Park (later Trafalgar Road), the most significant early street development to the south of the walls, also suggest the location of a significant, if declining, industrial workforce in the area. The return for 1841 records the presence of 11 masons, 4 quarrymen (the figure for Tenby as a whole was about 30)[58] and 21 labourers; that for 1861, 5 masons, 1 stone quarrier, 1 quarryman, 1 stone cutter, 10 labourers and 5 general labourers; that for 1881, 5 masons, 3 labourers and 4 general labourers; and that for 1901, 2 masons, 2 labourer-masons and 2 general labourers.[59] By the later nineteenth century, as the area to the west of the walled town was taken over for residential development, so it became increasingly de-industrialised; the dominant occupational grouping for Chimney Park in the 1881 census was the 22 laundresses recorded, reflecting the feminisation of the street and the growth of the service sector.

Figure 6.10 Lime kilns, Kiln Park, Tenby, 2010.

It is possible that the quarrymen and similar workers listed in the census returns for Chimney Park were employed on the Black Rock site, only a short walk across the Burrows. This was the location of a substantial complex of limestone quarries and limekilns.[60] The development of this major industrial site (situated now in Kiln Park Holiday Centre) is unclear, but it seems likely that a first block of seven kilns (see Figure 6.10) was constructed *c.* 1800 to service the upsurge in demand for agricultural lime at the time of the Napoleonic Wars.[61] At this stage, the River Ritec still flowed underneath the south cliff at Tenby and was incorporated into a tidal lake, Holloway Water, which lay to the rear of the Burrows (see Figures 6.2 and 6.11) and lapped against the quarries at Black Rock, allowing access for boats to bring in coal and culm and take out lime. In 1811, Sir John Owen built a wall, running from Black Rock to Tenby south cliff, to drain and reclaim the salt marsh to the rear of the Burrows, and though this was subsequently breached and had to be rebuilt, and it is likely that the area flooded frequently, it must have marked the end of easy water access to the kilns.[62] However, an even more impressive bank of 12 kilns – described as 'one of the most striking industrial monuments in the county' – was constructed in *c.* 1865, this time serviced by a branch line of the newly arrived Pembroke and Tenby Railway (opened in 1863).[63]

Figure 6.11 Tenby from Penally, looking across the Burrows and the golf links, towards South Cliff, with St Catherine's Island in the background to the right.

Around the time of the building of the new kilns, residential development had started on the western side of the rock on which Tenby stood. In the long term, this was to transform the character of this part of the town and the environs adjacent to it. The closure of the town quarries and their absorption into residential development, the opening of the golf course on the Burrows in the 1890s, the construction in 1929 of M. W. Shanly's Pavilion – known derogatively as 'Shanlyville on Sea' – across the original inlet of the River Ritec, the full incorporation of the South Sands with their two-mile long sweep from St Catherine's Island to Giltar Point into the suite of the town's bathing beaches, and the post–Second World War development of the Kiln Park caravan and camping facility on the site that included the Black Rock quarries and kilns marked the transformation from an industrial to recreational landscape. But change did not happen overnight. Transition was a protracted process. It would appear that thousands of tons of sand was taken from the South Sands and sold for industrial purposes in the early twentieth century, prompting some local criticism,[64] and the Black Rock quarries were reported as still 'active' in 1953, though it is unlikely that much limestone was being extracted by this time.[65]

Segregation and Incorporation

There is a tendency to view resorts as places wholly devoted to leisure and to focus on those parts of their makeup and construct historical narratives of creation or transition that reflect this. Guide literature is particularly prone to adopting this perspective since its market is the visitor population, for whom it supplies not only empirical data but also appropriate cultural images. Yet, all resorts have a basic economic structure – of food suppliers, transport services, building craftsmen, domestic servants, general labourers and the like – that is fundamental to their effective operation but is not unique to them. For those resorts that emerge out of an older port, the situation, as we have seen in the case of Tenby, is further complicated by the continuing presence of working maritime and industrial elements, so that any simple narrative of transition from port to resort tells only part of the story of the rise of the seaside town. This narrative is in addition complicated by the fact that Tenby displayed some of the characteristics of a county town. Although Haverfordwest was the assize town for Pembrokeshire, Tenby also serviced the consumer and recreational needs of the regional gentry, reflected in the presence of horse races, winter hunt week and the Tenby and County Club.[66] However, although the 'county' elements of the town's profile reinforced its programme of resort recreations and decorous image, the maritime and industrial elements had the potential to do the opposite. In 1890, a call for a larger local police force was prompted by a claim that on two occasions in June for two hours 'a great portion of the south side of the town was in the hands of the mob' when servicemen based in the vicinity 'came into collision with the civilians, largely made up of Devonshire and other fishermen who made Tenby their port of debarkation for their fish and use the harbour on a Saturday and Sunday for their weekly resort'.[67] There were two solutions to managing the clash between port and resort: segregation or incorporation.

The purely industrial elements were very largely incompatible with the life of a fashionable resort. Coal, quarries and kilns did not mix easily with parasols and promenades. The solution in the case of Tenby was simply to confine the two worlds to different zones, the coal mining to the north of the town, the quarrying and its operatives to the southern and western portions of the promontory and the adjacent area bordering on the Ritec estuary. It is true that the South Sands, the Burrows and Giltar Point were favourite locations for visitor excursions from the town but what is marked is how the guidebooks largely ignore the industrial undertakings to be found there.[68] Over the longer term, zoning became unnecessary as the area de-industrialised and, in the twentieth century, became enveloped in resort development.

The maritime elements posed different problems. Because the harbour was the centre of both port and resort, it was impossible to spatially separate the two functions. Although the Croft and the North Beach provided a measure of zoning, the presence of the early baths and assembly rooms, Castle Hill with its fashionable walks and the Castle Beach all in close proximity to the harbour, which was also the point of arrival and embarkation for many visitors, made segregation unviable. It was simply not feasible to insulate the visitors from the bustle, dirt and smell of a working port with sailors, quay men, fishermen and their families, who crowded into ramshackle cottages in or close to the harbour. Although, in theory, this threatened the polite ambience of the resort, in practice, it became possible to incorporate the experience of the port into that of the resort.

The sea, in all its manifestations, was part of what visiting a resort was about. In 1800, one guidebook commended how 'the distant fishing boats with their white sails, and the voices of fishermen, who constantly frequent this coast, borne at intervals in the air, are circumstances which animate the scene'.[69] This was an early sign of the fashionable cult that celebrated and aestheticised the working sea and those employed in its service and that reached its apogee in the later Victorian and Edwardian periods in movements such as the Breton and Cornish schools of painting.[70] In 1868, William Freeman, describing the animated scene on the beach, observed,

> A picturesque sight, but one you do not frequently see, is the shrimper, often a Welsh girl, with her flannel dress tucked up above her knees, wading through the flood and pushing her net before her. Then there are the Tenby fishermen, casting their nets for mackerel, which are often very plentiful. But the shrimper, with the water breaking round in white curling eddies, is decidedly the most attractive sight to an artistic eye – handkerchief tied over-head – a stray curl escaping behind, arms bared to the elbow, tight fitting check flannel dress gathered up behind over a red petticoat; in the distance she looks like a water nymph.[71]

The sexual overtones may not always have been present, but painters such as Edward Joseph Head and William Powell Frith produced images of Tenby fishermen and fisherwomen that extolled the virtues of work and the healthy, rugged, independent and primitive qualities of their subjects.[72] Not only did this play to the values of the middle class but also offered an antidote to its anxieties about the dangers of industrial life and in particular the degeneration of the urban working class.

The Frith picture of a Tenby fisherwoman (painted in 1880) was in fact of a Llangwm fish seller (a small village about 12 miles west of Tenby located on the Daugleddau waterway inland of Milford Haven), one of a group who regularly appeared in the town selling prawns and oysters. Images of these women, appearing individually or in groups, dressed in traditional costume and carrying their wicker baskets or 'creels', acquired an iconic status. Mason's guidebook reproduced a drawing of a Llangwm fisherwoman, accompanying it with a text describing how such women 'are natives of a village of that name ... and are of a race peculiar to that place' and how 'the Llangum [*sic*] people universally inter-marry, and thus the peculiarities of their race are perpetuated'.[73] There is more than a hint here that the visitors' interest in the fisherwomen in part arose from a quest for, and fascination with, a racially and genetically determined primitivism – akin to that to be discovered in the darker corners of the Asian and African continents, though in this case to be found on the almost equally benighted Celtic peripheries – that helped to define their own sense of civilisation.[74] The picture in Mason's guide was an engraving, but the Llangwm fisher-women were also widely photographed for commercial products such as postcards. One of the professional photographers who captured the fisherwomen (see Figure 6.12) was Charles Smith Allen (1831–1897), a pioneer of photography in Wales with a formidable output and reputation, whose 'Excelsior Studios' – operating in the mid to late Victorian period – occupied the old assembly rooms in Paxton's bath house and directly overlooked the harbour. It was a fortuitous location, allowing him to take many photographs of working fishing boats, fishermen, fisherwomen and their families, reflecting not only his own aesthetic predilections but also those of his public.[75]

At some point in the early twentieth century, it must have become clear that Tenby's days as a working port were numbered. This accelerated a trend that had been discernible for some time by which the town's maritime resources, capital and labour were being put at the service of its leisure regime. Recreational angling was one option open to visitors. In 1868, a guidebook observed, 'During the summer, fishing excursions are sources of amusement. The necessary lines, hooks, and bait may be hired of the boats' crew, and in the "mackerel season" this becomes an exciting sport'. For those seeking something more 'authentic', it was possible for 'gentlemen who delight in salt water experiences ... [to] arrange with the skippers of the fishing skiffs to accompany them in their fishing trips; provisions must be taken for the purpose, and the amateur will have to "rough it"'. The same guide referred to 'yachting and boating excursions. ... The circuit of the expansive and beautiful Bay of Carmarthen is frequently a favourite excursion by water ... The island of Caldy [*sic*] ... is the chief point of attraction to seaward, and in the summer months, during the Tenby season, it is every day visited by parties who pic-nic on its beach or cliffs'.[76] When Beatrix Potter visited Tenby in 1900 she wrote to a friend that 'it has been so hot lately, the only

Llangwm Fish Women

Figure 6.12 Llangwm fisherwoman, Charles Allen Smith (1831–1897).

cool place is on the water in a boat. I go out every morning and I generally tell the boat man to row close under the cliffs so that I can watch the birds'. It is likely that on occasions like these, local boats and men were being redeployed from commercial maritime occupations to meet the needs of the visitors. In 1935, 35 motor, sailing and rowing boats were licensed in Tenby to carry between 6 and 12 passengers.[77]

Over time, Tenby added, like many fashionable resorts, yachting to its package of pleasures, with an annual regatta by the late nineteenth century, if not before.[78] Late Victorian and Edwardian photographs of the harbour show what would appear to be some larger yachts and smaller pleasure craft.[79] As the commercial business of the harbour declined, so it would seem that its resources were turned to the leisure sector. In 1936, the 'Sailing Club' was renting the 'Hut on [the] Old Pier', and in the following year, J. W. Edwards applied for a vacant store on the harbour

> for the building and repairing of small boats ... I am at present engaged in building my third sailing boat in six months but find the matter of finding a suitable workshop an extremely difficult proposition. ... I would like to point out that a great deal of my work has been brought about by the activities of the Tenby Sailing Club's members whose enthusiasm, I believe, will eventually make their Club one of the attractions of Tenby.[80]

The long-term transformation of Tenby from port to resort is clear enough. It is a process that has affected not only the town but also its surrounding environs and the Pembrokeshire littoral as a whole. Few holidaymakers now visiting the bracelet of small resorts north of Tenby, between Saundersfoot and Amroth, would imagine that, less than a century ago, it was an active coalfield. One way of interpreting these changes is to construct a narrative of transition that emphasises outcomes and focuses on the progressive rise of the resort. However, as the case of Tenby testifies, this obscures a more complicated, protracted and multi-stranded process by which continuing (or new) maritime and industrial roles have, in the short and medium term, to be woven into the web of change. In this sense, it is more credible to talk not of a single narrative but of narratives of transition. The case of Tenby also emphasises the importance of, as the essays in this volume demonstrate, the special mix of circumstances surrounding each resort, so that the process of transition will vary considerably from place to place and, in some cases, will go into reversal, with the commercial and industrial elements re-emerging as powerful strands. This may prompt a reconsideration of how far in the nineteenth century it is accurate to see urbanisation as inextricably linked to differentiation and specialisation. The rise of the seaside resort had the capacity to complicate as well as simplify a town's economic, social and cultural profile.

Notes and References

1. I should like to thank the Board of Celtic Studies who funded the project on 'Resorts and ports: Swansea, Tenby and Aberystwyth, 1750–1914' from which this study has emerged; my co-directors of the project Louise Miskell and Owen Roberts; and Kate Sullivan, the research assistant on the project, who undertook bibliographic work and collected primary material. I am grateful for the considerable assistance I have received from the staff of the National Library of Wales, and from Mark Lewis, collection manager, and Sue Baldwin, the honorary librarian, at Tenby Museum.

2. David Palliser, ed., *The Cambridge urban history of Britain*, Volume I, *600–1540* (Cambridge, 2000), pp. 100, 469–70, 484, 712–13; Peter Clark, ed., *The Cambridge urban history of Britain*, Volume II, *1540–1840* (Cambridge, 2000), pp. 134–6; R. A. Griffiths, ed., *The boroughs of medieval Wales* (Cardiff, 1978); I. Soulsby, *The towns of medieval Wales* (Chichester, 1983), pp. 250–3; L. Owen, 'The population of Wales in the sixteenth and seventeenth centuries', *Transactions of the Honourable Society of Cymmrodorion* (1959), pp. 107–12; N. Powell, 'Do Numbers Count? Towns in early modern Wales', *Urban History*, 32 (2005), pp. 48–51.
3. Copy of a journal of a tour in Wales and England 1767–1768, National Library of Wales (NLW), MS 147C, fos. 27–32, 26–8 Sept. 1767.
4. An account of three tours by an English gentleman (1787–96), NLW, Deposit 9532A, 20 Aug. 1787, fo. 26; J. Baker, *A picturesque guide to the local beauties of Wales*, Volume I (London, 1791), p. 165; D. Howell, 'Society, 1660–1793', in B. Howells, ed., *Pembrokeshire county history*, Volume III, *Early modern Pembrokeshire, 1536–1815* (Haverfordwest, 1987), p. 263.
5. P. Borsay, 'Health and leisure resorts, 1700–1840', in Clark, *Cambridge urban history*, II, pp. 776–80; A. Durie, *Scotland for the holidays: tourism in Scotland, c. 1780–1939* (East Linton, 2003), pp. 66–9; J. K. Walton, *The English seaside resort: a social history, 1750–1914* (Leicester, 1983), pp. 11–16, 48–51; A. Brodie and G. Winter, *England's seaside resorts* (Swindon, 2007), pp. 11–26; J. F. Travis, *The rise of the Devon seaside resorts, 1750–1900* (Exeter, 1993), pp. 7–47.
6. *Aris's Birmingham Gazette*, 13 July 1763; *Hereford Journal*, 29 July 1771, 9 July 1772, 27 May 1779. I owe these references to the late David Lloyd. W. J. Lewis, *Born on a Perilous Rock: Aberystwyth Past and Present*, 3rd edn (Aberystwyth, 1980), pp. 194–9.
7. P. Borsay, 'Welsh seaside resorts: historiography, sources and themes', *Welsh History Review*, 24 (2008), pp. 92–119; N. Yates, *The Welsh seaside resorts: growth, decline, and survival*, University of Wales, Lampeter, Trivium Publications, Occasional Papers, 1 (2006), pp. 2–10; D. Boorman, *The Brighton of Wales: Swansea as a Fashionable Resort, c. 1780– c. 1830* (Swansea, 1986).
8. B. Howells, 'Land and people, 1536–1642', Idem, 'The economy, 1536–1642', Howell, 'Society, 1660–1793', in Howells, (ed.) *Pembrokeshire County History*, III, pp. 18–19, 85–7, 293–4; Soulsby, *Towns of medieval Wales* , pp. 252–3; M. R. Connop-Price, *Pembrokeshire the forgotten coalfield* (Ashbourne, 2004), p. 36.
9. NLW MS 147C, fo. 27, 27 Sept. 1767; C. Norris, *Etchings of Tenby* (London, 1812), p. 17.
10. Howells, 'Land and people', Howells (ed.) *Pembrokeshire County History*, III, p. 16; Soulsby, *Towns of medieval Wales*, p. 253.
11. A. Corbin, *The Lure of the sea*, translated J. Phelps (Cambridge, 1994).
12. NLW Cwrtmawr 393C, Walk through South Wales (1819), fo. 65; NLW Deposit 18943B, Narrative of a tour (late 18th/early 19thc?), fo. 19; NLW Deposits 6685C, Journal of a tour in Wales (1831), fo. 26, 19 Aug. 1831; 'Tour through South Wales and some of the adjacent English counties', in W. Mavor, ed., *The British tourist's companion, or traveller's pocket companion, through England, Wales, Scotland and Ireland* , V (London, 1798), p. 155.
13. *British parliamentary papers; 1851 census of Great Britain ... population 7*, reprinted (Shannon, 1970), p. 32; *British parliamentary papers: 1891 census of England and Wales ... volume II: population 22*, reprinted (Shannon, 1970), p. 1080; Walton, *English Seaside Resort*, p. 65.
14. Walton, *English Seaside Resort*, pp. 53–5, 58, 60–2, 65–6, 68–9.

15. Tenby Museum (TM), Borough of Tenby Order Book (1776–1835), fo. 21, 5 Oct. 1781; TM, TEM/SE/BOX 36, lease 17 Oct. 1805, between the mayor, bailiffs and burgesses of Tenby and Sir William Paxton.

16. *Carmarthen Journal*, 16 July 1810, 28 July 1810; *The Tenby guide* (Swansea, 1810), p. 17; H. C. Colvin, *A bibliographical dictionary of British architects, 1660–1840* , 3rd edn (New Haven and London, 1995), p. 263; J. Tipton, *Fair and fashionable Tenby; two hundred years as a seaside resort* (Tenby, 1987), pp. 11–12; T. Lloyd, J. Orbach and R. Scourfield, *The buildings of Wales: Pembrokeshire* (New Haven and London, 2004), pp. 469, 479, 481.

17. G. Nicholson, *The Cambrian traveller's guide* (London, 1813), p. 1276; J. Feltham, *A guide to all the watering and sea-bathing places*, new edn. (London, [1815]), p. 472; NLW Deposits 6685C, fo. 26, 19 Apr. 1831.

18. These can be tracked in the town's first dedicated guidebooks; *The Tenby guide* (1810); [C. Norris], *An account of Tenby* (Pembroke and Tenby, 1818); M. A. Bourne, *A guide to Tenby and its neighbourhood, historical and descriptive* (Carmarthen, 1843); George Henry Hough, *Sketches of Tenby and its neighbourhood* (Tenby, 1846); R. Mason, ed., *A guide to Tenby and its neighbourhood*, 5th edn (Tenby, [1867]); *Allen's guide to Tenby*, ed. F. P. Gwynne (Tenby, [c. 1868]).

19. TM, C. Norris, Tenby Sketchbook, copied by E. H. Edwards with notes by E. Laws (1903), fos. 19–20; Gwynne, *Sketches of Tenby*, p. 3; Lloyd et al., *Pembrokeshire*, pp. 480, 483; Tipton, *Fair and fashionable Tenby*, p. 16; Soulsby, *Towns of medieval Wales*, pp. 251–2.

20. 1 Vict., c. 13, An act for the improvement of the borough of Tenby (1838); Mason, *Guide to Tenby* [1867], pp. 18–19, 162, 188.

21. TM, TENBM (1983) 4256, Map of the town of Tenby from an actual survey in the year 1849, E. B. Hughes, surveyor, Narberth; Lloyd et al., *Pembrokeshire*, pp. 482–1. The only significant built areas to the south and west of the town walls, are a short stretch of Chimney Park and small knots of development around the Deer Park, the Green and in the Greenhill area.

22. TM, Tenby Corporation Minutes, 5 Jan. 1846.

23. R. Parker, *The Railways of Pembrokeshire* (Southampton, 2008), pp. 35–48.

24. *Allen's guide to Tenby* [c. 1868], p. 12.

25. TM, Tenby Corporation Minutes, 24 Aug, 23 Nov and 21 Dec 1868, 8 Feb. and 20 Apr. 1869; Tipton, *Fair and fashionable Tenby*, p. 26.

26. *Tenby Observer*, 29 June 1882, 24 Feb. 1887; Lloyd et al. *Pembrokeshire*, pp. 484–5; R. Lewis, 'The towns of Pembrokeshire, 1815–1974', in D. W. Howell, ed., *Pembrokeshire County History*, Volume IV, *Modern Pembrokeshire, 1815–1974* (Haverfordwest, 1993), pp. 50–1.

27. Tipton, *Fair and fashionable Tenby*, p. 28.

28. *Tenby Observer*, 29 March and 24 May 1877, 26 July 1894; *Allen's guide to Tenby* [c. 1868], pp. 139–41; A. L. Leach, *Guide to Tenby*, 3rd edn (Tenby, 1908), pp. 41–3; *Seaside watering places ... season 1900–1901* (London), p. 397; Tipton, *Fair and fashionable Tenby*, pp. 42–3.

29. NLW Deposits 6685C, fo. 26, 19 Aug. 1831; T. W. Hordley, *Reminiscences of T. W. Hordley* (1934), pp. 22–3; see also *Carmarthen Journal*, 23 Sept. 1842; Walton, *English seaside resort*, p. 77.

30. *Tourist's guide to Tenby and neighbourhood* (Tenby, 1881), p. 12 London: E. W. Allen.

31. Compare the visitors' lists in the *Tenby Observer* for July-September 1869 with those for July-September 1880; *Cambrian*, 21 and 26 July, 18 Aug., 15 and 22 Sept. 1871, 15 and 29 May, 23 and 31 July, 11 Sept. 1875, 7 July, 4 and 11 Aug., 15 Sept. 1876.

32. TM, TD/Hotels and Tourism, 5 June [1912], [J. P.] Evans on behalf of Aberavon Shop Assistants' Committee to [Tenby town clerk]; TM, TEM/SE/37, 25 Aug. [1898], F. F. Mazuchelli to [Tenby town clerk].

33. Bourne, *Guide to Tenby*, [Tipton, FFT, 19–20].

34. Mavor, *British tourist's companion*, V, p. 155; NLW Cwrtmawr 393C, fo. 65; NLW MS 1347B, 'Letters addressed by Charles Norris to the burgesses and inhabitants of Tenby July 9th 1831 also remarks upon the above letter and upon the town of Tenby ', fo. 75.

35. 1 Vict., c. 13.

36. *Carmarthen Journal*, 9 Sept 1842; Gwynne, *Sketches of Tenby*, pp. 72–3; TM, TEM/SE/36, Account book for the building of the new pier (1842–3).

37. TM, TEM/SE/35, (Draft) provisional order for the construction, maintenance and regulation of a pier at Tenby in the county of Pembroke (1893); TEM/SE/36, 24 Jan. 1898, Charles Dagnell and Co. to Tenby town clerk; TEM/SE/37, 22 Nov. 1869 and 7 Feb. 1873, reports from George Bowen, harbour master, to the Tenby corporation; National Archives Kew, MT10/604, 619, various papers of the Board of Trade, Harbour Department, relating to the new landing pier at Tenby (1892–3).

38. Norris, *Etchings of Tenby*, p. 19; the phraseology is repeated in S. Leigh, *Leigh's Guide to Wales and Monmouthshire* (London, 1831), p. 317.

39. A. H. Galvin, *Sea of change: 19th century maritime activity in S. E. Pembrokeshire* (Coventry, 2002), pp. 3–5; D. W. Howell, 'Farming in Pembrokeshire, 1815–1974', in Howell, *Pembrokeshire County History*, IV, p. 79.

40. Hordley, *Reminiscences*, pp. 13–14.

41. Galvin, *Sea of Change*, pp. 18–21; R. Howells, 'The Pembrokeshire islands', in Howell, *Pembrokeshire County History*, IV, p. 217; TM, TEM/SE/36, leases of storehouse on old pier, 14 Aug. 1913 and 14 Sept. 1916; TM, TR/5/1/3, Harbour rental 1911–40.

42. TM, PhB16 'Tenby Harbour and Royal Victoria Pier'.

43. For the history of the fishing industry in Tenby in the nineteenth century see Galvin, *Sea of Change*, pp. 53–91; K. McKay, 'The port of Milford: the fishing industry', in Howell, *Pembrokeshire County History*, IV, p. 175; M. Smylie, *The herring fisheries of Wales* (Llanwrst), 19980, pp. 15–22.

44. *The Tenby guide*, pp. 10; Norris, *Etchings of Tenby*, p, 63; Leigh, *Leigh's Guide to Wales and Monmouthshire*, p. 317.

45. C. F. Cliffe, *The Book of South Wales, the Bristol Channel, Monmouthshire, and the Wye* (London, 1847), pp. 225–6; Mason, *Guide to Tenby* [1867], p. 159.

46. Galvin, *Sea of Change*, p. 69.

47. Galvin, *Sea of Change*, pp. 74–5.

48. 1881 UK census collection, Ancestry.co.uk. Many of these families had disappeared by the 1901 census.

49. TM, TEM/SE/BOX 36, 6 May 1898, Thomas George to the Mayor and Gentlemen of the Corporation of Tenby.

50. [Norris], *Account of Tenby* (1818), p. 68; *Allen's guide to Tenby* [c. 1868], p. 3.

51. D. Howell, 'The economy, 1660–1793', in Howells, *Pembrokeshire county history*, III, pp. 319–25; B. John, *Pembrokeshire* (London, 1978), pp. 100–13.

52. M. R. Connop-Price, 'The Pembrokeshire coal industry', in Howell, *Pembrokeshire County History*, IV, pp. 111–37; Connop-Price, *Pembrokeshire the forgotten coalfield*; NLW MS 147C, fo. 27, 26 Sept. 1767; William Matthews, *Miscellaneous Companions* (1786), I, [Notes 84]

53. TM, TEM/SE/ 35, Tenby harbour papers 1726; Feltham, *A guide to all the watering and sea-bathing places*, p. 471; Galvin, *Sea of Change*, pp. 23–43;

Connop-Price, *Pembrokeshire the forgotten coalfield*, p. 173. Though Saunders-foot was the key port for the shipment of coal, it is interesting to note that a photograph exists of Tenby harbour in *c*. 1910 with a steamer apparently being loaded with culm (Connop-Price, *Pembrokeshire the forgotten coalfield*, plates between pp. 144–5), and that in the first decade of the twentieth century a coal exporter and shipbroker from Cardiff wrote to the town clerk stating that 'We have been offered Gas Coke f.o.b. Tenby, and should like to know whether there is any wharfage space where shipment could take place … If you would write to us, informing us whether shipping could be dealt with, principally Coke, loaded onto steamer at Tenby, how shipment takes place, whether from wagon over tip, or by basket, sack, etc. we should be extremely obliged' (TM, TEM/SE/37/45, 9 Nov 190?, S. Franklin Jones to Tenby Town Clerk).

54. Galvin, *Sea of Change*, pp. 44–52.

55. Norris, *Etchings of Tenby*, p. 62; Gwynne, *Sketches of Tenby*, pp. 63, 92–3; P. H. Gosse, *Tenby: a seaside holiday* (London, 1856), pp. 126, 270; J. G. Jenkins, *Maritime Heritage: the ships and seamen of southern Ceredigion* (Llandysul, 1982), p. 46; TM, Black Rock Quarries file; P. B. S. Davies, *Pembrokeshire limekilns: limekilns and lime burning around the Pembrokeshire coast* (St David's, 1997).

56. Ordnance Survey maps, 1": 41.66ft, Pembroke (Southampton, 1888–9), sheets XLI, 11.13–14; see also TM, Borough of Tenby Order Book, 7 May 1793, for the construction of a limekiln on the western side of Pill Fields. John, *Pembrokeshire*, p. 108, states that there were still 17 limekilns in Tenby in 1908.

57. *Allen's guide to Tenby* [c. 1868], p. 12; *Tenby Observer*, 23 Sept 1869, 6 Jan., 19 Feb. and 24 Feb. 1870.

58. M. B. Evans, 'The land and its people, 1815–1974', in Howell, *Pembrokeshire County History*, IV, p. 29.

59. 1841, 1861, 1881, 1901 UK census collection, Ancestry.co.uk..

60. T M, Black Rock Quarries file; Royal Commission on the Historic and Ancient Monuments of Wales, Cadw: Welsh Historic Mounments Ancient Monuments Record Form, PE436; Davies, *Pembrokeshire limekins*, p. 8.

61. R. Williams, *Limekilns and limeburning* (Princes Risborough), pp. 7–8; Davies, *Pembrokeshire limekins*, p. 22.

62. Norris, *Etchings of Tenby*, p. 29; TM, Norris, Tenby Sketchbook, fos. 4–7; TM, Egerton Allen Papers, N179/4, 'When Tenby's rivulet went underground'; Leach, *Guide to Tenby* (1901), pp. 14–15.

63. Lloyd et al., *Pembrokeshire*, p. 354; Parker, *Railways of Pembrokeshire*, pp. 35–48.

64. TM, TEM/SE/35, various documents and letters concerning removal of sands, 1899–1909.

65. T M, Black Rock Quarries file, fo. 12.

66. *Carmarthen Journal*, 15 Feb. 1839; *Tenby Observer*, 18 and 25 Jan., 29 Nov. and 6 Dec. 1877; D. W. Howell, 'Leisure and recreation, 1815–1974', in Howell, *Pembrokeshire county history*, IV, p. 421; Tipton, *Fair and fashionable Tenby*, pp. 20–1, 28, 35; R. Lawrence, *The rise and fall of Tenby races, 1846–1936* (Brecon, 2003).

67. *Tenby Observer*, 19 June 1890.

68. See, for example, Mason, *Guide to Tenby* [1867], pp. 35–9, 48–53; Leach, *Guide to Tenby* (1901), pp. 14–15, 50–8.

69. *The Cambrian directory, or, cursory sketches of the Welsh territories* (Salisbury, 1800), p. 38.

70. C. Payne, *Where the sea meets the land: artists on the coast in nineteenth-century Britain* (Bristol, 2007), pp. 171–99; T. Cross, *The shining sands: artists in Newlyn and St Ives 188–1930* (Tiverton, 1999).
71. W. Freeman, *My summer holiday: being a tourist's jottings about Tenby* (London, 1868), p. 48.
72. *Edward J. Head: Resident Tenby Artist, 1863–1937* (Tenby, 1998); N. Westerman, *Edward Joseph Head, the Tenby Artist; a biography* (Slough, 2001).
73. Mason, *Guide to Tenby* [1867], pp 143–4; *Allen's guide to Tenby* [c. 1868], pp. 135–6.
74. M. G. H. Pittock, *Celtic identity and the British image* (Manchester, 1999), pp. 20–93.
75. R. Worsley, *Prince of places:a pictorial and social history of Tenby featuring the work of pioneer photographer Charles Smith Allen* (Haverfordwest, 1979); TM, PhB 17, Tenby beaches and Castle Hill.
76. *Allen's guide to Tenby* [c. 1868], pp. 97–9, 106, 108.
77. Beatrix Potter. *The 'Tenby' Letters* (Tenby, n.d.), 24 April 1900, Beatrix Potter to Marjory; TM, TEM/SE/35, List of pleasure boats and licensees (1930).
78. J. Cusack, 'The rise of yachting in England and South Devon revisited, 1640–1827', in S. Fisher, ed., *Recreation and the sea* (Exeter, 1997), pp. 101–49; Galvin, *Sea of change*, pp. 177–86.
79. TM, PhB 16, Tenby Harbour and Royal Victoria Pier.
80. TM, TR/5/1/3, Harbour Rental 1911–40; TEM/SE/BOX 36, 22 April 1937, J. W. Edwards to Meyrick Price.

Chapter 7

A Town Divided? Sea-Bathing, Dock-Building and Oyster-Fishing in Nineteenth-Century Swansea

LOUISE MISKELL

Introduction

The decades of the late eighteenth and early nineteenth centuries in Britain are associated by historians of tourism and leisure with the growth in popularity of sea bathing as a health and leisure activity among the fashionable elite.[1] By economic historians, they are viewed as decades of industrial take-off, when the pace of output from textile manufactories, smelting works and mines quickened to unprecedented levels.[2] In south Wales, as elsewhere, interest in these two historical developments has tended to run along different parallels. The region has been recognised by British resort historians as an early sea-bathing destination,[3] but traditionally, historians of the economy and industry have provided much the more dominant view of its development. As a region rich in natural coal and iron ore deposits, south Wales' coastal towns are better known for their functions as commercial ports than for their attractions as tourist destinations. The shipping of coal from harbours and inlets along the coastline stimulated urban development from west to east as Llanelli, Neath and Swansea, Newport, Cardiff and Barry all felt the effects of escalating levels of mining activity inland, in the coalfield valleys spanning the counties of Carmarthenshire, Glamorgan and Monmouthshire.[4] One consequence of this is that industrial perspectives of Welsh urbanisation have been brought to the fore, and places which quite patently developed both tourist and trading functions have been either neglected or misunderstood. This chapter re-evaluates the history of one south Wales coastal town by merging these two previously separate strands of research to form a new analysis of interactivity between industry and tourism.

The town of Swansea, in the county of Glamorgan, occupies a broad sweep of bay on the Bristol Channel coast, at the mouth of the River Tawe, some 40 miles to the west of Cardiff. Along with the English ports of Southampton and Tynemouth, it has been identified as a town where, by early Victorian times, 'resort functions were fading in relative importance alongside the development of heavy industry and docks'.[5] The building of commercial docks in the nineteenth century, as far as seaside historians are concerned, diverted coastal towns such as these away from becoming tourist centres, with the two roles being seen as incompatible. At Tynemouth, the sea–beach had been an important area of recreation and public gathering,[6] and in Southampton, spa gardens were laid out in the second half of the eighteenth century as the town developed as a bathing resort and watering place, but these fell into disuse by the second decade of the nineteenth century as the town's resort attractions were eclipsed by other rising tourist centres on the English south coast and as the town developed as a port.[7] This chronology of sea-bathing functions giving way to industry by the middle of the nineteenth century has also been embraced in Swansea, in local studies of both the metal industries and of sea bathing. In particular, the construction of Swansea's South Dock, between 1852 and 1859, is identified as a key period in 'Swansea's decline as a fashionable resort and its development as a commercial port and industrial town'.[8] Historian of the copper industry, Ronald Rees, described Swansea as 'a town divided in body and spirit' as the growth of smelting works began to overwhelm the Georgian bathing resort in the early decades of the nineteenth century.[9] Meanwhile, Swansea's resort historian, David Boorman, concluded that the town's success as a fashionable resort in the late eighteenth and early nineteenth centuries was a 'passing phase', which ended with the decision, in the 1840s, to construct new commercial docks on the foreshore.[10]

This simple model of resort decline in the face of commercial dock development has an attractive symmetry, but it will be argued here that a different explanatory framework is required in order to understand the relationship between industry and leisure in Swansea, one which allows for the possibility that sea bathers and other users of the beach were capable of adapting to the commercial encroachments on their patch and which acknowledges a more multi-faceted relationship between the town and its coastline. Tourist hubs relocated along the coast as traditional bathing areas were swallowed up by successive phases of dock building. In common with many other British seaside resorts of the period, Swansea's tourist clientele began to diversify in the second half of the nineteenth century, with more day-trippers and working-class visitors from the surrounding industrial suburbs and further afield. In addition, attempts were made to retain the reputation of Swansea Bay as a destination for wealthier visitors. Not only was there a continuous history of sea bathing in nineteenth-century Swansea, the dialogue between industry and tourism in the town was also continuous. As the tourists moved west to newly developing bathing areas, away from the docks, so debates over commercial and leisure uses of the coastline moved with them. Far from existing in separate and mutually exclusive worlds, tourism and industry in Swansea were forced to interact because of their equal dependence on the town's coastal spaces. The result was an evolving discourse on the relationship between the town and the bay, accessible to the historian in guidebooks, newspaper columns and visitor comments of the day and crucial to understanding the urban environment of Swansea.

The Development of Industry and Seaside Tourism, *c.* 1790–1840

The emergence of Swansea as a world centre for copper smelting has been the dominant theme in histories of the town in the modern era, earning it the nickname 'Copperopolis'.[11] With the copper ore fields of Cornwall, a relatively short sea journey away across the Bristol Channel, and abundant seams of suitable coal deposited all along the coast, the town was ideally situated to develop as a smelting centre. These factors, along with the advantages of a large natural harbour, led to the establishment of the first smelting works in the area in 1717 by a Bristol entrepreneur, Dr John Lane.[12] Other investors soon followed, from London, the English Midlands, Cornwall and North Wales so that, by the beginning of the nineteenth century, there were seven smelting works in operation on the banks of the Tawe River.[13] But, far from dominated by copper, Swansea's economic portfolio in the period from 1700 was diverse. In the early decades of the eighteenth century, Swansea traded with French, Spanish and Portuguese ports, all of which imported coal, grain and woollen goods carried from Swansea.[14] In addition, the early years of the eighteenth century were prosperous ones for the western end of the south Wales coalfield. The bituminous coal found in this area was a highly favoured fuel for both industrial and domestic purposes, valued for its easy lighting and slow-burning qualities.[15] Landowners in the area around Swansea, and west towards Gower, Lougher, the Burry estuary and Llanelli, were aware of the profits to be made from the discovery of coal on their lands, and investment in exploration and mining activity was encouraged. The produce of their collieries met demand for coal in Devon, Somerset, Cornwall and Ireland. Alongside this industrial activity there was, by the early nineteenth century, a range of other economic activity including two earthenware manufactories; a developing retail, financial and commercial sector; and a thriving trade in health tourism, with wealthy visitors flocking in during the summer months to bathe in the sheltered bay.[16]

In the 1789–1815 period, the South-West was emerging as Britain's 'second major concentration of early coastal resorts' (the first being the English south coast).[17] The Welsh side of the Bristol Channel played its part in this development, and Swansea's significance was noted early by visiting writers. In the 1790s, James Baker recorded that the town 'had a very considerable share of resort from the most distinguished persons of fashion in the kingdom; it is found a most convenient trip for the inhabitants of Bristol, Bath and the counties adjoining the Severn Sea'.[18] Like the Devon resorts of Torquay, Sidmouth, Lynton and Lynmouth, Swansea found that it had the potential to cater for a new demand for sea bathing among the social elite, and the corporation did its best to capitalise by investing in the best available facilities for visiting bathers. They took as their yardstick the resorts of the English south coast, and in 1789, a representative was dispatched to Weymouth to see the latest bathing 'machines' in operation and to 'procure from hence an exact model of one of the best machines made by measure'.[19] The Duke of Beaufort, Swansea's principal landowner, added further momentum by creating a public walk on an area of recently enclosed corporation land lying between the town and the sea known as 'the burrows'. This became the town's first real visitor hub with a 'pleasant promenade', 'many good lodging houses' and 'two convenient bathing-houses'.[20]

Members of the local business community threw their weight behind similar initiatives. Local entrepreneur George Haynes took the lead in establishing a weekly English language newspaper, the *Cambrian*, in 1804. The new publication placed Swansea on a par with other resort towns where newspapers served as a useful organ of the tourist trade, announcing the arrival of the well-heeled and fashionable and informing visitors of the events and services on offer throughout the season.[21] Likewise, the gentry of the town and neighbourhood initiated a subscription to fund the building of new assembly rooms and a theatre for use by 'fashionable strangers'[22] but, as one local guidebook pointed out, with the added advantage of serving their own recreational demands.[23] Along with the opening of several libraries, these improvements helped to attract tourists to early nineteenth-century Swansea and to convince guidebook writers that the town was as serious about its resort functions as it was about its industrial development.

The drive to lure fashionable tourists at the same time as expanding as a centre for copper smelting was not unproblematic. One guidebook author while describing Swansea as 'a favourite resort in the summer for bathing', also warned that 'the volumes of smoke from the different manufactories are a great deduction to the general attraction of the place'.[24] The copper smoke emitted from the smelting works contained high levels of sulphur, which was both unpleasant to smell and toxic to breathe. Proprietors of the copper works found themselves involved in litigation with claimants seeking compensation for the damaging effects of the smoke.[25] But, carried inland from the copper works, which were themselves situated a couple of miles upriver from the bay and the fashionable Swansea sands, it was the lower Swansea Valley rather than the coastal strip which bore the brunt of industrial pollution.[26] The copper smoke did not drive the sea bathers away. In fact, tourism in

Swansea derived some indirect benefits from industry in the town in the late eighteenth and early nineteenth centuries. The Swansea to Oyster-mouth tramroad, for example, established in 1804 for the conveyance of limestone, iron ore and coal, was soon providing an excursion for visitors who were keen to travel in the horse-drawn carriage on rails, which ran along the seashore towards Mumbles. One passenger wrote approvingly of 'the scenery being grand, particularly Oystermouth Castle and the bay of Swansea'.[27] Local guidebooks directed visitors to the town's manu-facturing premises as well as the more natural appeal of its bay and sands. The 1802 *Swansea Guide* presented the proximity of bathing facilities and commercial sites as a positive advantage, drawing visitors' attention, for example, to the Cambrian pottery operated by George Haynes, which was arranged 'on Mr Wedgwood's plan', and situated 'contiguous to ... [his] Cold and Hot Sea Water Baths'.[28]

Dock-building and the Move to the West, c. 1840–1870

Although the sea bathers who visited Swansea in the early nineteenth century were not put off by the haze of copper smoke which hung about the periphery of the town, a much more serious challenge to their activities was posed by the building of commercial docks on the area of the foreshore traditionally used for bathing and promenading. On 26 September 1859, the *Morning Chronicle* carried a report, abridged from a special supplement printed a few days earlier in the *Cambrian* newspaper, of the opening of the new South Dock in Swansea. The description of the dock emphasised its size: 'The works comprise a spacious trumpet-mouth entrance, a half-tide basin, an immense lock, an iron bridge, and an inner dock of sufficient acre to allow some hundreds of ships to repose majestically on their shadows in perfect safety.' The report also conveyed the impact of a special branch line, which linked the new dock to the South Wales Railway, bisecting some of the main pedestrian and vehicle routes between the town and the bay:

> Wind Street is crossed by an iron bridge, and the line passes along towards the Royal Institution, where another iron bridge spans the main thoroughfare leading to Fisher Street and down to the Burrows. From this point its course is through Burrows Lodge-grounds where the arches terminate.[29]

Although it swallowed up the burrows, and thereby encroached directly on tourist territory, there were relatively few voices raised in the town in opposition to the dock development. The total number of vessels entering Swansea harbour each year had increased from 2295 at the end of the Napoleonic Wars to 3699 vessels by 1835. The 1830s and '40s also saw an expansion in overseas trade as Swansea smelters began to source copper from overseas ore fields.[30] All of this commercial activity put pressure on the harbour, the limitations of which were particularly problematic for larger ships. Despite some modernisation and improvement work under the harbour acts of 1791, 1796 and 1804, Swansea was still equipped with only a basic tidal harbour where vessels were liable to experience considerable delays at low tide.[31] Improvements were vital to sustain the prosperity of the local coal and copper industries whose representatives feared the commercial headway being made at Cardiff and Newport where docks had recently been constructed. That their wishes prevailed was not merely an example of commercial interests outweighing the needs of visitors. The town's preparedness to sacrifice the burrows seems also to have been based on a belief that tourism would not be stifled as a result. There was an awareness in mid-nineteenth-century Swansea that, with or without the docks, patterns of coastal tourism were changing. One local guidebook of the period noted that 'since the advent of the South Wales Railway, the resort of sea-bathing people has been very considerable'.[32]

In common with the expansion of the railway network to other British coastal towns, the extension of the south Wales line westwards via Swansea and into Carmarthenshire made new parts of the Welsh coast accessible to day-trippers as well as long-stay holidaymakers for the first time. Swansea was able to cater to these visitors, even after constructing new docks, thanks to its wide expanse of bay, which offered space for recreation to the west of the old burrows area. It was along the bay, at a point known locally as the 'slip', where a new sea-bathing hub emerged. The slip began to feature prominently in the lists of Swansea attractions highlighted by guidebook authors. It was described as 'a spot ... admirably suited for bathing' served by 'a large number of bathing machines ... drawn in and out of the water by horses', where visitors could 'breathe the invigorating breezes of the sea and enjoy the sight of the rippling wave and angry billows, or the surging and foaming breakers dashing against the shore'.[33] From the 1870s, the residents of Swansea's outlying industrial suburbs were able to access the slip via a new tramway system, which gave them a cheap and easy means of movement from their homes to the coastline. Passengers from Morriston, for example, could disembark at St Helen's station from where they could visit the beach or the nearby Victoria Park, which opened in 1887.[34] As new railway lines were constructed, the accessibility of the coast increased. The Swansea Bay to Rhondda line, built in 1895 to improve access to the growing coalfield, provided a convenient passenger service to the coast for the district's workers. Contemporaries anticipated that 'before the summer of 1895 has deepened into autumn, the teeming thousands of the Rhondda will have shown by their presence in our streets, our places of business and our pleasure haunts, that Swansea has practically added a wing to its municipality.'[35]

As well as catering for local day-trippers and working-class visitors from the surrounding industrial valleys, promoters of tourism in Swansea did not forget their more traditional client base among the social elite. Just as the South Dock construction project was reaching completion, the Swansea-based publishing firm of Pearse and Brown went into production with a series of guidebooks encouraging visitors to explore the most westerly reaches of Swansea Bay, around Mumbles.[36] Mumbles was a fishing community, consisting of 3574 residents and 742 houses in 1871, located some four miles west along the coast from Swansea.[37] Although well placed to capitalise on the westward drift of seaside tourism, which followed the mid-century dock developments in Swansea, it was oyster fishing which formed the most important part of its economy for much of the nineteenth century. At its height, in the early 1870s, some 188 oyster boats were engaged in dredging off Mumbles Head, employing over 600 people offshore and on land and yielding at least 10 million oysters per year.[38] During this period, guidebooks unequivocally placed the summer tourist trade behind the local fishing industry as the mainstay of the Mumbles economy but, at the same time, they point to its ability to attract wealthy seaside tourists. Local civil servant Leopold Charles Martin, who published a new guide to *Swansea and Gower with the Mumbles and Adjacent Bays* in 1879, described Oystermouth and Mumbles as 'a long straggling village skirting the western point of the charming bay of Swansea' where, 'although a much frequented and fashionable bathing place in the summer, little attempt has been made to render it more than a fishing, oyster dredging, and Lighthouse station'.[39]

Martin's observation that Mumbles was already an established destination for fashionable tourists is confirmed in the pages of local newspapers and directories. Two hotels were in operation there by the late 1850s and a further 28 Mumbles residents were offering residential accommodation in the form of lodgings by the same date.[40] In the 1860s, the *Cambrian* was publishing lists of 'Arrivals at the Mumbles', just as it had recorded the fashionable visitors flocking to Swansea in the summer season in earlier decades. The roll calls of names and addresses showed a strong Wales and West Country representation and a desire to publicise the fashionable and respectable characteristics of the visitors. In early October 1869, the arrivals included

the Misses Fortune Powell, staying at the Bath House Hotel; Miss Jones of Cardigan; Mr D. W. Pugh, Hereford; the Misses Gulley, West town, Somerset, staying at Caswell Hotel; Mr John Allen and Mr E. Hyde, at the George Hotel; E. M. Schmidt, Clifton, at Langland; F. T. Palgrave esq. and Mrs Palgrave; Sir A. H. Elton Bart. and Lady

Elton at Rosehill; Rev. R. Price, Mrs Price and family, Brecon; and Miss Price, Hagley, Herefordshire, at the Bath House Hotel.[41]

Seaside Tourism and the Fishing Industry, *c.* 1870–1900

For much of the 1860s and 1870s, the fishing and tourist trades seem to have co-existed quite happily in Mumbles. The oyster-fishing season began in September, thus avoiding the most popular months for sea bathing. It also serviced the needs of visitors by supplying unique take-home mementos of the area in the form of oysters, 'stewed and pickled and placed in jars, large quantities being taken away during the summer season by visitors, or sent away as appropriate souvenirs of the neighbourhood in which the fish are caught'.[42] Those visitors who did arrive during the fishing season were sometimes encouraged to add the industry to their itinerary of sightseeing. Willison's 1907 guide claimed that 'the storing and packing of the fish is very interesting, as also are the auction sales, which are held every morning. But here a word of warning; do not risk the fish market in clothes that are new or easily spotted, or much of the pleasure of the visit will be lost'.[43] By following such advice, holidaymakers in Victorian Mumbles were by no means unique. As Jan Hein Furnée shows elsewhere in this volume, as early as the 1660s, visitors to the Dutch seaside at Scheveningen were being encouraged to view the local fish auctions,[44] and by the end of the nineteenth century, there was, as Peter Borsay's chapter shows, a 'fashionable cult' of the sea as a place of work.[45] Although a little different from the fascination he identifies with the Pembrokeshire fisherwomen, the studio portrait of a blind Mumbles oyster fisherman, taken in Swansea in 1874, can be viewed in a similar vein. Thomas Brookman, pictured barefoot, in his ragged working clothes, with a net slung over his shoulder and basket at his waist, must have presented an image of working life which appeared to the well-to-do Mumbles tourist to be remote and other-worldly.[46]

Perhaps inevitably, as Fred Gray demonstrates in his essay on Brighton, there were some direct conflicts of interest between fishing and tourism. The oyster 'perches' occupied valuable foreshore space between high- and low-tide marks, where the oysters were deposited prior to being sent to market, and further out, immature oysters were left to develop in specially designated areas known as 'plantations'.[47] These arrangements were not easily compatible with sea-bathing activities and, with some late-season visitors arriving in September and October,[48] there was the potential for conflict between their interests and those of the oyster fishing community. But serious aggravation was averted because of a shift in the relationship between resort functions and fishing in Mumbles. Oyster fishing in the area was in decline from the mid-1870s. Just as the demise in the shipping industry headed off conflict between the tourists and maritime residents of Sørlandet, Norway, as shown in Berit Eide Johnsen's chapter,[49] so in Mumbles, tourism gained the upper hand as over-exploitation, lack of investment and, above all, pollution from sewage outlets and associated health concerns all contributed to depleting catches.[50] But the decline of the Mumbles oyster fishery was probably also hastened by the increase in visitor traffic. The extension of transport facilities for visitors added to the problems of the oyster-fishing community with the extension of the Mumbles railway to the pier in 1895, which removed the traditional 'lay-up' area used by fishing vessels in the bay.[51] The momentum had swung decisively in favour of the development of the area as a visitor destination in the 1880s, influenced perhaps by Gladstone's visit in June 1887. The Prime Minister, in the company of Swansea industrialist and Liberal MP, Henry Hussey Vivian, undertook a sightseeing tour of Mumbles bay and sampled a local oyster bar and temperance tavern.[52] The visit of such a distinguished figure seems to have galvanised local interest in the development of visitor facilities in the area and to have reinforced the perception of Mumbles as a resort for the social elite. A feature of contemporary comment about the development of visitor facilities in the area in the 1880s was the absence of any effort to appeal to the new mass market for working-class seaside tourism. Summer amusements at Mumbles, one *Cambrian* correspondent wrote, ought to emulate 'what prevails on the Continent at Baden Baden and other fashionable resorts; that is to say, that facilities for boating, lawn tennis, croquet, cricket, rinking, whist and chess playing, swimming and a musical band should be provided'.[53] Guide-book authors clearly expected late nineteenth-century visitors to the area to have the time and the financial means to stay for long enough to undertake extensive explorations of the wider coastline. The *Excelsior Guide* to Swansea and Mumbles published in 1880 directed them to the Gower coast and the bays of Langland, Pwlldu and Oxwich, as well as the Bishopston valley, the castles at Pennard and Penrice, and Worm's Head.[54] It was a pitch designed to appeal to the adventurous excursionist with resources and flexibility rather than to the casual day-tripper.

Conclusion

This evidence from the western reaches of Swansea Bay in the latter decades of the nineteenth century suggests that there were strands of continuity in the patterns of seaside tourism from the Georgian era. The coming of the working-class day-tripper did not extinguish Swansea's aspiration to attract wealthy holidaymakers, although it did so with more limited success in the period after 1880 than it had in the early years of the nineteenth century. In common with the experience of Tenby, examined elsewhere in this volume, competition from more exclusive resorts in other parts of the United Kingdom and on the continent meant that the task of attracting the wealthy tourist would always be more difficult in the late nineteenth century. This was despite the best efforts of the town council to respond to changing market trends, for instance with its new guidebook titled, *Swansea: The Naples of Wales*, published in 1925.[55] Dock building certainly disrupted patterns of seaside tourism in Swansea, but it did not bring them to an end. In common with a number of the towns further east on the Welsh side of the Bristol Channel, geographical factors meant that the needs of tourists could still be accommodated. The movement west of sea-bathing areas in Swansea in some respects anticipated the development of separate docks and seaside space in the late nineteenth-century at Barry and Penarth. The south-western-facing sands of Barry Island and the long seafront at Penarth became dedicated seaside leisure areas, distinct from the commercial dock areas, which grew up around the inlet of Barry Sound and the estuary of the River Ely.[56]

At Swansea, the prospect of dock expansion continued to loom large in debates over coastal development. An underlying concern that docks might engulf the remaining recreational areas of the coastline can be detected in newspaper comments and guidebook literature alike. The latter could barely keep pace with the rate at which the coastline was being developed for commercial shipping. L.C. Martin's 1879 guide, for example, was produced just as a new phase of dock construction was commencing. An additional page detailing its progress was inserted just before the book went to press, giving details of the new 'East Dock' that was due to occupy another 23 acres of the bay at Swansea.[57] Against this background, news of 'improvement' initiatives around Mumbles harbour was treated with suspicion. In January 1888, a *Cambrian* editorial on 'The Future of Mumbles' praised the construction of a promenade but queried the wisdom of a railway extension and pier building project at Mumbles Head:

> If ... there is the slightest intention of transforming the Mumbles into a coal-shipping station, then that is a very different thing and one which demands careful consideration ... It is perfectly clear that, while the Mumbles might well and wisely be improved as a watering place, and might easily provide for an increased passenger traffic, it would be folly to attempt to erect docks or shipping piers there while the accommodation in Swansea Harbour not more than two miles off, is not anything like fully utilised.[58]

Such sentiments revealed the degree to which contemporaries were still sensitive to the competing claims on their coastline. The relocation of the sea-bathing facilities after the dock-building phase of the mid-nineteenth century had not resulted in a settled vision of tourism and commerce occupying separate territory in the town. Instead, negotiation of the uses of the bay was an ongoing feature of urban life in Swansea in the nineteenth and early twentieth centuries.

It has not been the purpose of this chapter to present Swansea as a 'typical' example of how the relationship between seaside leisure and commerce was negotiated in the nineteenth century. Many of the circumstances affecting Swansea may not have been present in other coastal communities even if, as the other contributions to this volume show, there are parallels to be drawn with ports, fishing communities and industrial centres elsewhere in Europe. What the Swansea case does demonstrate is the value of considering issues of resort and port development over the long term. By tracing the ebb and flow of industry, trade and leisure patterns in the town from the 1780s to the 1920s, it is apparent that the balance between resort and port functions see-sawed. Analysis based on single-event turning points, such as the opening of a new dock or the effect of a new railway line, provides only a momentary glimpse of a complex picture deserving of a more lingering view.

Notes and References

1. See, for example, A. J. Durie. 'Medicine, health and economic development: promoting spa and seaside resorts in Scotland, c.1750-1830', *Medical History*, 47 (2003), pp. 195–216.
2. For example, T. S. Ashton, *The industrial revolution, 1760-1830* (London; New York, 1948).
3. P. Borsay, 'Health and leisure resorts, 1700–1870', in P. Clark, ed., *The Cambridge urban history of Britain, volume II, 1540–1840* (Cambridge, 2000), p. 778.
4. For example, M.V. Symons, *Coal mining in the Llanelli area, volume I: sixteenth century to 1829* (Llanelli, 1979); C. D. J. Trott, 'Coal mining in the borough of Neath in the seventeenth and early eighteenth centuries', *Morgannwg*, 13 (1969); R. Rees, *The black mystery. Coal mining in south-west Wales* (Talybont, 2008); M. J. Daunton, *Coal metropolis: Cardiff, 1870-1914* (Leicester, 1977); J. W. Dawson, *Commerce and customs. A history of the ports of Newport and Caerleon* (Newport, 1932).
5. J. K. Walton, *The English seaside resort. A social history, 1750–1914* (Leicester, 1983), p. 47.
6. See for example, the open air lecture held there in 1838 on the occasion of the BAAS's visit to Newcastle. An account of this is quoted in O. J. R. Howarth, *The British Association for the Advancement of Science: a retrospect, 1831–1931* (London, 1931), pp.101–2.
7. E. M. Sandell, 'Georgian Southampton: A watering-place and spa', in J. B. Morgan and P. Peberdy, eds., *Collected essays on Southampton* (Southampton, 1958), pp. 80, 86–7.
8. J. A. Owen, *Swansea's earliest open spaces. A study of Swansea's parks and their promoters in the nineteenth century* (Swansea, 1995), p.16.
9. R. Rees, *King copper. South Wales and the copper trade, 1584–1895* (Cardiff, 2000), p. 19.

10. D. Boorman, The *Brighton of Wales. Swansea as a fashionable seaside resort, c.1780-1830* (Swansea, 1986), p. 95.

11. See for example, S. Hughes, *Copperopolis. Landscapes of the early modern industrial period in Swansea* (Aberystwyth, 2000); E. Newell, '"Copperopolis": the rise and fall of the copper industry in the Swansea district, 1826–1921', *Business History*, 32 (1990), pp. 75–97.

12. R. O. Roberts, 'Dr John Lane and the foundation of the non-ferrous metal industries in the Swansea valley', *Gower*, 4 (1951), pp. 18–24.

13. R.O. Roberts, 'The smelting of non-ferrous metals since 1750', in G. Williams and A. H. John, eds., *Glamorgan county history volume V: industrial Glamorgan* (Cardiff, 1980).

14. For details see D. T. Williams, 'The Port Books of Swansea and Neath, 1709–1719', *Archaeologia Cambrensis*, 95 (1940), pp. 203–205.

15. Symons, *Coal mining in the Llanelli area*, pp.11-12.

16. For details see L. Miskell, *Intelligent town: an urban history of Swansea, c.1780–1855* (Cardiff, 2006), pp. 86–87.

17. Borsay, 'Health and leisure resorts', p.778.

18. J. Baker, *A Picturesque guide to the local beauties of Wales; interspersed with the most interesting objects of antiquity in that Principality* (2nd edn, London, 1791), pp. 126–27.

19. West Glamorgan Archives Service (WGAS), B/S Corp B8, Hall Day minute book, 26 October 1789.

20. *Mathews Swansea Directory* (Bristol, 1816), p. 6.

21. P. Clark and R. A. Houston, 'Culture and leisure, 1700-1840', in Clark, *Cambridge uuban history*, II, pp. 595–6.

22. WGAS, Royal Institution of South Wales, George Grant Francis collection, GGF4 p.67, Swansea Tontine Society list of subsciber to the Swansea theatre, n.d. (c.1804).

23. *New Swansea Guide* (Swansea, 1823), p. 50.

24. G. Nicholson, *The Cambrian Traveller's Guide* (Stourport, 1808), p. 602.

25. R. Rees, 'The Great Copper Trials', *History Today*, 43 (1993), pp. 38–44.

26. Miskell, *Intelligent Town*, pp. 82-83.

27. T. Lloyd, 'The diary of a visitor to Swansea and Gower in 1821', *Gower*, 34 (1983), pp. 12–13.

28. *The Swansea Guide* (Swansea, 1802), pp. 1–12.

29. *Morning Chronicle*, 26 September 1859.

30. D. Boorman, 'The City and the Channel', in R. A. Griffiths, ed., *The City of Swansea. Challenges and Change* (Stroud, 1990), p.107.

31. Ibid., pp. 132–5.

32. J. Lewis, *The Swansea Guide, 1851* (Swansea, 1851), p. 12.

33. T. E. Bath, *Excelsior Guide to Swansea and the Mumbles* (Swansea, 1880), p. 26.

34. Owen, *Swansea's earliest open spaces*, pp. 50–9.

35. WGAS, D/DZ 445/2, scrapbook cutting relating to the completion of the Swansea to Rhondda railway, c.1895, with a retrospective report, 14 March 1895. Newspaper unknown, but possibly *Cambria Daily Leader*.

36. *Our village wants a church: a few words about the Mumbles and its surrounding bays, with sketches of some of the prettiest spots in aid of a subscription for a new church* (Pearse and Brown: Swansea, 1859); *Guide to the Mumbles and Adjacent Bays* (Pearse and Brown: Swansea, 1860); *Guide to the Mumbles and adjacent bays, containing and a graphic and descriptive account of the many beauties of the locality* (Pearse and Brown: 2nd edn, revised and corrected with additions, Swansea, 1862).

37. Census figures for Mumbles from 1821 to 1881 are cited in N. L. Thomas, *Of Swansea West. the Mumbles past and present* (Llandysul, 1978), p. 25.
38. C. Matheson, 'The oyster fishery at Mumbles, Glamorgan', *Transactions of the Cardiff Naturalists' Society*, 66 (1934), pp. 83–84.
39. L. C. Martin, *Swansea and Gower with the Mumbles and adjacent bays. A guide and handbook for visitors and tourists* (Swansea, 1879), p. 35.
40. *Slater's Royal National Commercial Directory and Topography* (Manchester, London, 1858), see section on 'Swansea, with the villages of Mumbles, Sketty, Morriston, Llangefelach, Llansamlet and Neighbourhoods'.
41. *Cambrian*, 8 October 1869.
42. *Guide to the Mumbles and adjacent bays* (Pearse and Brown, 1860), quoted in G. R. Orrin, ed., *James Orrin (1828-93): his life and times in Victorian Mumbles: a documentary history* (Taunton, 2008), p. 21.
43. T. H. Willison, *Swansea: the ideal place for pleasure, health and holidays* (Swansea, 1906), pp. 39, 47.
44. J. H. Furnée, 'A Dutch idyll? Scheveningen as a seaside resort, fishing village and port, c.1700-1900', p. xxx.
45. P. Borsay, 'From port to resort: Tenby and narratives of transition, 1760–1914', p. xxx.
46. J. Andrews, 'A blind fisherman' (Swansea, 1874). This image is the property of Swansea Museum. A digitised version can be viewed at 'Gathering the Jewels', www.gtj.org.uk/GTJ70460.
47. R. J. H. Lloyd, 'The Mumbles oyster skiffs', *Mariner's Mirror*, 40 (1954), p. 263.
48. The *Cambrian* newspaper reported visitor arrivals at Mumbles throughout October in 1869.
49. B. E. Johnsen, 'Recycled maritime culture and landscape. Various aspects of the adaptation of nineteenth-century shipping and fishing industries to twentieth-century tourism in southern Norway', p. xxx.
50. G. Parsons, 'The uses and abuses of scientific expertise in English inshore oyster fishery, 1860–1910', in K. R. Benson and P. F. Rehbock (eds), *Oceanographic History. The Pacific and Beyond* (Washington, 2002), pp.400–401.
51. Lloyd, 'The Mumbles oyster skiffs', pp. 264–5.
52. An account of this can be found in Thomas, *Of Swansea West*, pp. 27–28.
53. *Cambrian*, 4 May 1888.
54. Bath, *Excelsior Guide*, pp. 58–61.
55. *Swansea. The Naples of Wales* (Swansea, 1925), p. 5.
56. B. Luxton, 'Ambition, vice and virtue: social life, 1884–1914', in D. Moore, ed., *Barry. The Centenary Book* (Barry, 1984), pp.310–13. For Penarth, see *Kelly's Directory of Monmouthshire and South Wales* (1895), pp.586–87.
57. Martin, *Swansea and Gower*, pp. 16–17.
58. *Cambrian*, 27 January 1888.

Chapter 8

Port and Resort: Symbiosis and Conflict in 'Old Whitby', England, since 1880

JOHN K. WALTON

A common theme in the history of seaside tourism is the way in which an ailing maritime economy provides cheap accommodation and a picturesque environment to enable and encourage the development of a tourist industry, alongside and within existing economic and social arrangements. Ports, particularly fishing ports, offered an exciting 'otherness', access to a different and distinctive way of life, which quickly became attractive to artists during the well-documented changes in attitudes to marine environments from the eighteenth century onwards, generating an aesthetic of the seaport and the 'fishing quarter' that reinforced its attractiveness to tourists who visited the coast for health and pleasure. Tensions between fishermen and fashionable visitors over desirable stretches of coastline are as old as coastal resorts, as the early removal of fishing nets from the Steine promenade at Brighton demonstrates[1]; but conflicts sharpened when an aesthetic of the 'modern' seaside, healthy, clean, rational and up-to-date, emerged at the turn of the nineteenth and twentieth centuries, gathering momentum during the 1920s and 1930s alongside persisting celebrations of the old, traditional, informal and scruffy. At this time, the fishing port turned resort encountered conflicts at the points of contact between old and new, and the authenticity of fishing communities was challenged in some places by the modernisation of boats and practices and also by engagement with the new economic opportunities of the tourist season. Proposals to demolish fishing quarters as slums and health hazards, to be replaced by new developments on the cliffs above the unplanned and unsanitary environment around the fishing beach or harbour, generated sustained conflict, particularly between the 1930s and the 1960s, as part of more general patterns of tension between tradition and modernity as seaside resorts embraced the new pleasures and aesthetics of sunshine, bodily display, athleticism, streamlining, concrete and curves.[2] This chapter explores these issues in twentieth-century Whitby, North Yorkshire, England.

Whitby might seem easy to dismiss as remote, provincial, stagnant and of little interest outside its own locality. It might be assimilated to standard assumptions about the inexorable decline of the traditional British seaside resort since the 1970s in the face of competition from new holiday destinations and patterns of consumption, fashion and desire.[3] Even Beatty and Fothergill's revision of the bleakest assumptions about the 'seaside economy' over the last 30 years of the twentieth century offered little comfort for Whitby itself. Its population at a spring census, not quite 13,000 in 2001, perpetuated more than a century of stagnation alternating with gentle decline. Its low-season claimant unemployment rate in January 2002 placed it third among the 43 largest British seaside resorts, two-thirds higher than the national figure for this kind of town. Taking hidden unemployment into account, only five other coastal resorts were worse off than Whitby, in spite of recorded growth of 29% in employment between 1971 and 2001, which put the town 13th in a similar league table. But, although most seaside resorts were also net importers of migrants of working age, another indication of relative economic health, Whitby propped up this table with a net loss of 8% over the 30 years.[4]

These depressing economic and demographic indicators hide a contrasting dimension to Whitby's current fortunes. It has enhanced its reputation as a popular weekend destination for seekers after distinctiveness and authenticity. Its day-tripper markets from old industrial areas of northeast England and West Yorkshire have been augmented by touring coach parties and more distant affluent visitors, keeping tourism buoyant through times which were harder in other resorts. In 2006, it won the title 'Best Seaside Resort' from the British consumer magazine *Holiday Which?* Newspaper publicity was supportive, referring to the town's 550,000 visitors per year, tourism employment running at one in five of the population, sandy beaches, quaint harbour, abbey ruins, picturesque cliffs, fossils, jet ornament manufacture, smoked herrings ('kippers'), folk festival, regatta and literary and historic associations. The regional *Yorkshire Post* emphasised a considerable recovery from high unemployment rates in the mid-1990s, with extensive new investment, particularly around the harbour. The dissonance between Whitby's gloomy economic and demographic record and its positive image and trans-class popularity as a seaside resort suggests that Beatty and Fothergill's statistics represent only part of the story.[5]

Whitby's success was part of a wider renaissance of small British seaside resorts with character and atmosphere, the bearers of attractive traditions from the history of the English seaside. The other resorts in the *Holiday Which?* top 10 were all small, distinctive and difficult to access from major population centres. Several were not only remote, particularly to the London media, but also had close connections with steam railways, a particularly evocative 'heritage' attraction in Britain.[6] Small resorts with eccentricity and the capacity for catering to niche markets were of the essence, connecting with nostalgia for a secure, rich and interesting past which could be transmitted to a new generation by reviving an idealised family holiday in a traditional and evocative seaside destination. The quaintness associated with fishing industries and declining harbours was a great asset here.[7]

Whitby was well placed to capitalise on these preferences. It was hard to reach but not inaccessible, with a meandering railway branch line and winding, hilly road access across moors that challenged without deterring. Close to (and connected to) a preserved steam railway, it displayed a line of beach huts under the West Cliff, reached by an Art Deco cliff lift. Such assets ticked relevant boxes, as did the long sandy beach itself and a plenitude of rock pools, fossils, shells and marine fauna for secure children's play, exploration and education. But its key attractions were its topography, architecture, patina of history and aura of legend, and the ambience these attributes generated.[8] Whitby's tourist success can be largely ascribed to its membership of a particularly attractive category: the historic seaside resort with literary and artistic associations, from Caedmon to Mrs Gaskell and Bram Stoker. This depended heavily, though far from exclusively (the abbey on the cliffs was a great draw), on its maritime associations and fishing industry and therefore on the town's historic negotiation of its roles as port and resort

The positive side of Whitby's dual identity as an ailing urban economy and successful seaside resort is based on presenting and marketing its history, architecture and atmosphere to visitors who seek quaintness, customs and authenticity. Whitby's lack of economic and demographic dynamism enabled it to preserve enough desirable characteristics to offer its seductive combination of the standard attractions of a seaside resort with those of a historic maritime town offering small, specialist shops in narrow, crowded, atmospheric streets. This emerged from sustained conflict between resort interests and other elements of Whitby's economy during the 1930s and 1950s and between alternative visions of the future of the resort. A sequence of battles pitted advocates of modernisation, who urged the demolition of many of the old buildings by the harbour, against preservationists, who recognised the value for tourist purposes of the courts, alleys and stairways which climbed the valley sides above the harbour in an eccentric intricacy that delighted some as much as it affronted others. This chapter traces the relationships between Whitby's maritime past and tourist present, particularly its image since the late nineteenth century as a romantic, historic, quaint place to visit, and analyses key mid-twentieth-century conflicts, which had lasting and defining consequences, between advocates of preservation, modernity and compromise.[9]

Whitby was among the first British seaside resorts, but its attractive distinguishing features had other origins. Its economic heyday came in the late eighteenth century and first half of the nineteenth century, when wealthy ship owners benefited from the coal trade between northeast England and London and from buoyant overseas trade. It participated in the north Atlantic whale fishery, sustained an inshore and middle-water fishing fleet, established a prosperous shipbuilding industry and remained a significant commercial port. The architectural legacy includes the opulent Georgian houses for merchants, ship owners and master mariners that climb the hill from the western side of the harbour (an attractively unorthodox variant on the Georgian terrace heritage theme); the diminutive Georgian market hall and surviving cottages on the more plebeian East Side of the River Esk; the stone piers and lighthouses of the 1820s which protect the harbour entrance; and the neat classical station of the Whitby and Pickering Railway, which opened as early as 1836.[10] Whitby also had much deeper historical roots, with the abbey ruins and the engagingly eccentric parish church dominating the East Cliff (see Figure 8.1). There was a local fishing industry from at least the fourteenth century. A rural hinterland of heather moors and deep valleys added romantic landscapes and stories to those associated with the cliffs, the sea and the abbey, reinforced from the mid-1880s by local author Mary

Figure 8.1 Whitby from the air looking northwards in 1935, just before the first demolitions began, showing the estuary and harbour, with the abbey on the right above the packed housing on the steep slopes of the East Side.

Linskill.[11] The town also benefited from the mid-Victorian vogue for ornaments carved and polished from local jet, which contributed to the construction of an attractive traditional artisan identity.[12] Such attributes became staple themes of Whitby's Victorian guidebooks.

More important to Whitby's reputation as a quaint, old-fashioned settlement in which 'history' was accessible, tangible and democratic, an identification based on the emotions rather than the formal cultural capital of detailed historical and architectural knowledge, were the narrow streets, yards and stairways of the 'old town' on either side of the harbour, particularly the East Side, where (in favoured metaphors) warrens and honeycombs of houses were built up the valley sides, piled one on top of another in memorable confusion during the eighteenth and nineteenth centuries, giving visitors a strong sense of 'living history', a continuum between past and present (see Figure 8.2).[13] This aspect of Whitby was already generating positive reactions, and publicity, from George du Maurier and a coterie of metropolitan artists and contributors to *Punch* from the 1860s onwards, although most Victorian tourist and holiday guides had little or nothing to say about it.[14] Black in 1862 dismissed the harbour area thus: 'The streets are generally narrow, and the older parts of the town present nothing remarkable'. On the other hand, the West Cliff, the focal point of the new seaside resort at the top of the hill, 'has many very handsome buildings, affording excellent accommodation to visitors'.[15] The bulk of the text presented the abbey, parish church and surroundings, and the West Cliff.[16] At the end of the nineteenth century, the national coastal resort guidebook *Seaside Watering Places* remarked that 'the streets of Whitby are very narrow, and the houses quaint in appearance', without going into detail, but it added that 'millions of herrings are landed weekly ... the visitor should walk along the quay and observe the sale by auction, landing, salting, barrelling and despatch of the fish'.[17] This was an important aspect of the fishing port's contribution to the distinctiveness of Whitby as a tourist destination, but the herring fishery occupied only a few weeks, which happened to overlap with the holiday season, and its attractive liveliness came from the transient passage of vessels (and herring lassies) from Scotland (particularly) and Cornwall, as they followed the shoals down the coast. It also declined sharply during the inter-war years, to enjoy a brief revival after the Second World War.[18] Even without the herring, however, the regular fish market was a perennial attraction, as represented (typically) by Clive Rouse in 1936: 'The busy scene on the quay when the morning fishmarket takes place, with the masts of the fishing vessels masses (*sic*) in the harbour close by, is always attractive'.[19]

As late as 1934, Horne's local guidebook was still ignoring the quaint and antiquarian aspects of Whitby's appeal, with substantial sections on local legends and customs and the work of Mary Linskill but nothing

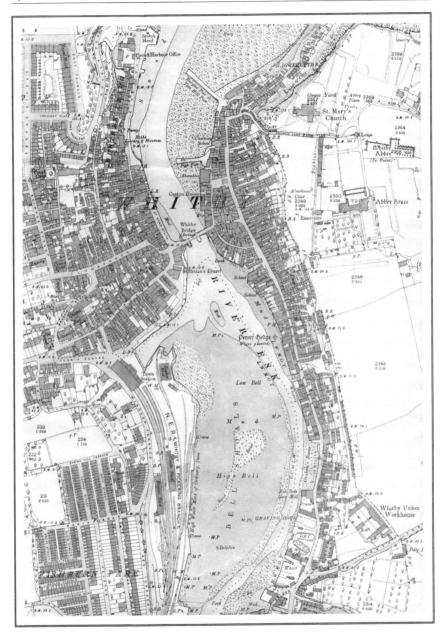

Figure 8.2 The 25-inch Ordnance Survey map of Whitby in 1895, providing a vivid picture of the cramped and tangled housing on both sides of the harbour.

about the streets and vistas of 'Old Whitby'.[20] Sir Nikolaus Pevsner recorded nothing of architectural interest in Church Street, the heart of Old Whitby, for his *Buildings of England* series in 1966, but he acknowledged that the town was 'delightful in three ways at least: for the abbey, which, as a ruin, is sublime as well as picturesque; for the parish church, the like of which is not to be found anywhere in the country; and for the busy quaysides with the long irregular rows of houses, picturesque in so different a way from the picturesqueness of the abbey'.[21] From at least the 1880s, this irregularity and informality of layout, with the activity of the fishing and commercial ports, was central to Whitby's attractions for growing numbers of visitors, reinforced by artistic representations of Old Whitby and its people through paintings and photographs, in spite of the unwillingness of some guidebooks to acknowledge this aspect of its appeal and of local hostility to the old town in some quarters. Indeed, the negative aspects of Pevsner's verdict may have been influenced by demolitions since the mid-1930s, as we shall see.[22] The seaside resort, as such, had developed away from the harbour, expressing the standard assumptions of the Victorian family holiday, developing the new accoutrements of the inter-war years (cliff lifts, art deco cafes, outdoor sporting facilities, a paddling pool), while offering nothing to stimulate the historical imagination.[23]

Following a common British seaside pattern, as at Brighton and Margate in the mid-eighteenth century, Whitby developed tourism as a counterweight to decline in the economic activities on which the town's Georgian and early Victorian prosperity had been based.[24] This invariably gave rise to tensions between the older activities of port and harbour and the new requirements of the holiday industry, although in Whitby these were muted for many years by physical segregation between the tourist accommodation district and its dedicated amenities on and beneath the West Cliff and the older harbour area in the Esk Valley, which was a place to pass through and inspect with interest while visiting the shops just off the harbour or walking through to the abbey and parish church on the East Cliff. The Victorian commercial port, which took advantage of the proximity of the railway station and locomotive depot to expand earlier facilities against the unpretentious backdrop of the new terraces of Fishburn Park, was not part of this picture, and from the 1950s onwards, its decline and replacement by car parking and retail use was relatively uncontroversial in tourism terms. 'Old Whitby', to use the conventional shorthand, was a different matter. It could be constructed as something attractively rather than threateningly 'other', the object of the romantic or at least the picturesque gaze, at a time when the quaintness and picturesque qualities of such places were becoming established, particularly (but not exclusively) among middle- and upper-class visitors with artistic or bohemian pretensions, wherever old fishing settlements were developing resort functions.[25] At Whitby, these unofficial and informal perceptions were increasingly important to the town's popularity as a tourist destination.[26]

This process began in earnest in Whitby and the district from the 1880s, with the development of artists' colonies in and around the nearby fishing villages of Staithes, Runswick Bay and Robin Hood's Bay, a common phenomenon on British and European coastlines during this period, as Peter Borsay notes in the case of Tenby.[27] It was reinforced by the popularity of the novelist Mary Linskill, who gave Whitby its romantic label of 'the Haven under the Hill' and turned its red roofs into an enduring symbol of refuge, comfort and hospitality, and the evocative photography of Frank Meadow Sutcliffe, who (like many of the artists of the Staithes Group) moved here from the industrial West Riding of Yorkshire. He was particularly successful at capturing harbour scenes and groups of fishermen and their families, who were willing to co-operate in preserving the illusion of spontaneity through the stillness required by a long exposure. By 1890, the local trade directory listed four artists and seven photographers (Sutcliffe included, alongside four fossil dealers and three antique furniture dealers) in Whitby at a time of economic transition. Seventy-five master mariners, 14 ship-owners, 5 boat builders and 2 shipbuilders were enumerated alongside 168 lodging-house keepers, 9 pleasure boat owners and a bathing-machine proprietor. Twenty-seven fishermen were listed as owners or part-owners of boats, 10 of whom lived in the old tenement and cottage areas of the west side of the harbour, 16 in the Church Street area of the East Side, and only 1 in the Victorian terraces of Fishburn Park, above the railway station. The perception of Old Whitby as romantic and attractive was enhanced by proximity to the harbour and association with the fishing industry and community, which conjured up virtuous images of courage, risk and hard work and associations with the Royal Navy, the lifeboat service and Britain as an island and seafaring nation. Fishing families were a small minority among the inhabitants of the West and (particularly) the East Sides of the harbour, but their presence coloured perceptions of the whole area. Sutcliffe's pictures of townscapes, maritime landscapes and local people became associated with a 'spirit of place' identified with tradition, character and authenticity, directing the tourist gaze towards such images and their immediate context, not least through the picture postcard market at the turn of the century, and heightening the attractiveness of the fishing quarter as a place to visit.[28]

Positive reactions to Old Whitby were propagated in the late nineteenth century, when the Society for the Protection of Ancient Buildings (founded 1877) began its campaigns against the destructive 'restoration' of medieval churches, and opposition arose in Whitby's regional hinterland to threats to medieval religious buildings in Leeds and York.[29] James Russell Lowell, United States minister to Britain between 1880 and 1885, visited Whitby regularly, enjoying it as a 'very primitive place' which reminded him of an earlier, more simple United States, and wrote to friends, 'I wish you could see the yards, steep flights of stone steps hurrying down from the West Cliff and the East, between which the river... crawls into the sea'.[30] A North Eastern Railway guidebook of 1904 emphasised Whitby's 'ancient and picturesque character' and its appeal to artists, particularly when fishermen animated the quay in the early morning.[31] Such perceptions were now commonplace, and when Sutcliffe, who had furthered them, retired from his photographic business in 1922, he was looked after by the local artistic and antiquarian establishment, becoming curator of the Whitby Museum. Whitby's place in the canon of romantic topographical and naturalistic maritime painting was illustrated by the *Whitby Gazette* at the end of 1936:

> From time immemorial Whitby has been one of the most attractive beauty spots around the English coast for all members of the painting fraternity, and seldom does a year pass but one finds pictures of it on the walls of the Royal Academy or other London exhibitions.[32]

For 'from time immemorial', we should read 'over the past half century', but the point was made. Three years earlier, just as controversy was breaking out over demolition and 'improvement' proposals for the old town, J.S. Miller of London provided a conventional eulogy of the romantic appearance of Old Whitby from the West Cliff: 'The smile of Whitby viewed from the Khyber Pass on a sunny day, when the light beats through the haze of smoke that seems to hang perpetually over these crazy, red-roofed cottages clustered up the sides of the East Cliff like the seeds in a ripe pomegranate'.[33] Nor was this just the province of the self-consciously 'cultivated' practitioner of artistic appreciation, as was indicated by a manuscript letter written by A. Halliwell of Bury after listening to a radio broadcast on the preservation or demolition of Old Whitby: 'I think that visitors to Whitby would agree with me that the medevial (*sic*) atmosphere of the place is its greatest attraction ... although I am no artist but just an ordinary working man I can apreciate (*sic*) the beauty of your town ... and would not like to see any of it disturbed'.[34] The importance attached to vistas and atmosphere in such commentaries, particularly those that suggest that these discourses had become as conventional as the eighteenth-century vocabulary of the picturesque (while adapting and partially incorporating it), is crucial to understanding the attraction exerted by this unassuming but compelling environment on its visiting public.[35]

G. Douglas Bolton's eulogy of Old Whitby, published in 1955, pulls all the strands together:

A gleam is apt to come into the eye of almost any artist when considering the pictorial attractions of Whitby. No other seaside town or fishing village can surpass it for old-world charm ... Best of all, the modern promenade and paths (on the West Cliff) command an unsurpassed view of the East Cliff, a scene of incomparable beauty when there is a fine sunset. At such times the old houses of the East Cliff reflect the setting sun, and glow with a rich emphasis on red pantiles and the irregular roof-lines of the cottages ... we look across the gleaming harbour, resplendent with sunset reflections, to the old houses with windows shining like burnished gold and red pantiles climbing towards the venerable walls of St Hilda's Abbey ... Whitby's past is rich and fragrant, mellow with age and spiced with the salt tang of the sea ... We start our exploration of Whitby from the West Cliff, where one overlooks the harbour, St Mary's Church, the abbey, the swing-bridge, the river estuary, and hundreds of small red houses clinging – almost by their pantiles – to the East Cliff like mussels to a rock ... The real fascination of Whitby begins once we are across the Esk ... Exploration of the "ghauts" or alleys can go on for hours. We are now in a land crowded with children; some are maybe dirty, dishevelled and mischievous but they are all friendly. We see fisherfolk in characteristic poses just waiting to be photographed, craftsmen discovered by chance in odd corners, red cottages nodding together, and exciting old houses with an atmosphere of the sea ... We next turn down the Tin Ghaut, a short alley leading to the estuary, where artists have gathered for generations. I stood there the whole of one July morning just to watch and photograph the changing scene ... I was not joined by any other artist or photographer, and so I gradually became a part of the scenery. After the first flutter of curiosity the local population became quite used to me, the children dispersed, the cats came out and purred, and washing was hung out on the waterfront just as if I was not there. Within three hours I had obtained an entire set of photographs and made friends with the local residents ... Eventually I was invited inside Fair Isle Cottage and shown around; spotless inside and out, this old-world building would have cheered the heart of Captain Cook.[36]

Even as this version of the tourist gaze, with all its overtones of internal orientalism, sentimentalism and the celebration of an exotic imagined 'other', became more influential and more important to the local economy, an alternative discourse of denigration was developing in parallel, seeing dirt, untidiness, squalor and disease where others were drawn to the quaint, the organic, the complex and the weathered. An emergent planning orthodoxy favoured comprehensive redevelopment, regarding inconveniently sited old buildings as obstacles rather than challenges and attracting popular support.[37] Housing reformers in early twentieth-century Whitby saw the labyrinthine alleys and yards above the harbour as irrational, unhealthy and dangerous blots on the landscape, ripe for clearance and replacement. In 1907, a local government board inspector used the language of 'slums' in attacking overcrowded and insanitary housing, denouncing the local authority for spending on holiday attractions rather than sanitary improvements. His report provided ammunition when housing became a live local issue after the First World War.[38] Whitby's medical officer of health argued that improvements were being made, that many houses were not bad enough to condemn but likely to become so if neglected, and that infant mortality remained below the national average. His 1914 report advocated some compulsory purchase and demolition but explained that sanitary reform was under way through the abolition of open channel drainage, the conversion of privies to water closets and the extension of a modern sewerage system.[39] Criticisms of housing conditions in Old Whitby gathered strength during the post-war housing crisis. In 1919, a letter to the local press claimed that conditions were worse than in London's East End, with piled-up tenements, water running down walls and no fresh air or sunlight in the houses. The 'unhealthy houses (should be) pulled down, the streets widened, and ... Whitby shall be made into a place of health and beauty ... Then will be the time for artists to paint Whitby with its red-tiled roofs that are watertight'.[40] This awareness of the artistic (and tourist) dimension to the question haunted subsequent debates. 'Owd Native' argued that the real East End slums were nothing like airy Whitby, with its clean and virtuous inhabitants, low death rates in the old town and 'fine specimens of humanity' who were brought up there, and pointed out that if the old streets and alleys were demolished, 'Whitby would soon be marked off the artist's programme'.[41] References to red roofs as justification for preservation, alluding to the sentimental imagery associated with Mary Linskill, aroused the ire of housing reformers who repudiated nostalgia and emotion and took pride in the scientific purity of their statistical analysis.

The question then faded into the background as the local authority retreated from building new council houses on cost grounds. Towards the end of the 1920s, a new programme of drainage improvements and water-closet installations began to make headway on the East Side. Meanwhile, in 1928 the Rector of Whitby visited the Ministry of Health in London to draw attention to the deplorable condition of the town's housing stock.[42] The medical officer of health's report of 1929 recorded that although only 10 of the 361 houses inspected in that year were 'unfit for human habitation', a further 220 were 'not in all respects reasonably fit' and a plea for a belated municipal house-building programme recognised that 'the existing conditions are a menace to the health of the occupants'.[43]

Conflicts on these lines occupied most of the 1930s, as the advent of a new medical officer of health, Dr Dale Wood, coincided with the election of a group of reformers to the Urban District Council, led by Kenneth McNeil, a boot and shoe dealer and crusader for clearance and redevelopment. This development acquired added significance when central government began, through the Housing Acts of 1930 and particularly 1933, to push a slum clearance agenda which threatened the wholesale destruction of Old Whitby. The Council responded to the collapse of several houses on the East Side in 1930 by inaugurating a small municipal house-building programme, but a central government circular of May 1933 precipitated further activity. It required a survey of the town's housing stock and a full report on the extent of the need for demolition, reconditioning and re-housing. The medical officer of health quickly completed what a Ministry of Health inspector described as a 'model report', but its proposals for the demolition of 450 houses, three-quarters of them on the East Side, aroused a storm of controversy within the town which soon spread across a wide area of northern and midland England.[44]

Debate within Whitby reached a climax when the Urban District Council rejected the report in October 1933, despite McNeil's efforts. Much discussion concerned the rights of property-owners, the unfairness or otherwise of evicting people from cheap housing to which they were attached and which was not demonstrably unhealthy and the cost to local taxpayers of demolition and replacement housing, but aesthetic and tourism issues were never far from the surface. In the local press, 'A Common Bricklayer' complained that the rhetoric of Old Whitby was being used to protect slum landlords and perpetuate the exploitation of tenants living in unacceptable conditions, while 'Light and Air' asserted that 'the persons ... for whom I have the most contempt are those poisonous sentimentalists who persist in moaning over the proposed demolition of "the old red roofs of Whitby." Through some of these roofs the rain trickles, sometimes freely', while rents were unacceptably high.[45] McNeil argued that conditions on the East Side actually repelled tourists, claiming that, in Church Street, 'over and over again he met visitors who turned back in disgust at the conditions they saw'. The medical officer of health emphasised the unacceptability of current conditions while recognising the problems posed by wholesale demolition:

> The red roofs of Whitby, which are so well known, and have for many years been an attraction for artists and other admirers, in

reality cover some very unsavoury conditions, which are far from
beautiful ... without any preconceived plan ... It is difficult to deal
with such conditions in any hope of bringing them into conformity
with present-day requirements, whilst, on the other hand, it is
unthinkable to clear away a whole town.[46]

As the report became more widely publicised, external opinion
opposed the demolitions. Mrs Janet King of Bedford Park, London,
denied that there were slums in Whitby. She had never spoken to a single
visitor who was disgusted by the East Side and had encountered nothing
but 'admiration expressed for the old place, so beautifully kept'. The
proposals would destroy 'the most picturesque seaside town in the
whole of England'. Bertha Doyle of Boston Spa near Leeds, a minor artist,
argued that visitors returned to Whitby because of its beauty and charm
and that '... real modern Progress preserves and treasures the heritage
and beauty of the past', such as the red tiles, 'so cheerful in our grey
climate, and so marvellously glowing at sunset time'. She pointed out the
dangers of creating all seaside resorts in the same modernised concrete
image and denied any health risk now that the old town's sanitation had
been dealt with. Whitby was 'an ancient monument, and should be
jealously guarded as such'. Rowland H. Hill of nearby Hinderwell,
another artist, emphasised the 'character and colour for which Whitby
has become so deservedly famous ... Both on aesthetic and commercial
grounds it would be a suicidal policy to mar this beauty in haste'. He
emphasised 'a widespread loyalty to its picturesqueness' among many
regular visitors, a use of the word that underscored its changing usage
since the eighteenth century and the broadening definition of what
constituted an attractive landscape and pictorial composition.[47] Thus
began a long negotiation, as local interests within Whitby, for a variety of
motives, sought to reduce the number and impact of the demolitions and
(in some cases) to preserve the character and atmosphere of the town, for
its own sake and as an attraction for tourists. Divisions on the Urban
District Council paralysed all attempts to designate redevelopment areas,
as efforts were made to reduce the number of affected houses. No single
programme commanded agreement, and Councillor Campion was one of
several to deny the relevance of experience elsewhere to the special case
of Whitby. A few small demolitions were agreed on the West Side in 1936,
but an impasse had been reached.[48]

The Council now sought advice from the Royal Institute of British
Architects, who recommended E.C. Bewlay, a Birmingham architect with
artistic interests and experience of Whitby. The Council agreed to hire
him as a consultant at the considerable fee of 300 guineas plus expenses:
a remarkable decision for a parsimonious local authority, demonstrating
the importance attached to the issue.[49] Bewlay's brief was to reconcile the

sanitary and the 'picturesque' by recommending what could be kept or reconditioned, what must be demolished, and how the lost buildings could be replaced with minimum damage to the character, atmosphere and townscape of Old Whitby. His report in August 1936 provided a new figure of 229 dwellings to be demolished (alternative proposals had ranged as low as 150), praised the cleanliness of the inhabitants and the impact of recent improvements and paid particular attention to preserving the pictorial quality of the urban landscape, particularly as seen from the cliffs and harbour bridge. He emphasised that rebuilding, where appropriate, 'should take place so as to retain unspoilt as far as possible the collective beauty of clustered roofs which is Whitby, and which no Town Planning Acts will ever allow to be repeated'. Detailed reports on particular areas paid heed to whether roofs and clusters of housing were visible at a distance and affected the visual experience of the onlooker. He hoped to achieve ' . . . an increase rather than a decrease in the beauty of the unrivalled views from across the river', as well as provide new accommodation for fishing families, in keeping with their surroundings.[50] The *Whitby Gazette* reacted favourably, recognising that a literal-minded statistical approach would bring needless destruction of 'buildings that by their delightful grouping and setting were a joy to anyone possessing any sense of beauty'. The attitudes of central government had recently become more sympathetic to preserving such urban environments.[51] The Whitby campaigners benefited from similar resistance in the fashionable Cornish resort of St Ives, where the artists' colony was particularly influential and well connected. The Bewlay report helped to shift debate against wholesale demolition, while Bewlay's architectural practice remained influential in designing new and replacement housing.[52]

As a result of all this and the interest taken by the influential Society for the Protection of Ancient Buildings and the Council for the Preservation of Rural England (CPRE, established 1926), debate on particular proposals for demolition and renewal was strongly coloured by questions of urban aesthetics.[53] The CPRE's Whitby branch was inaugurated in November 1936, and Vivian Seaton Gray, clerk to the Council, reported on discussion at the organisation's Torquay conference: ' . . . the individuality of resorts should be preserved as far as possible – a matter of particular importance to them at Whitby, that they should preserve the particular atmosphere they had at the present time . . .'.[54] When formal inquiries began into specific demolition proposals in 1937, however, Seaton Gray found himself in his official capacity supporting Council policy, while solicitors and builders speaking for owners and tenants borrowed the rhetoric of preservation and the picturesque in the hope of saving their homes and investments. When New Way Ghaut, on the East Side, was considered, Mr Graham, for the Council, argued that 'from a congestion point of view, it would not be possible to reconvert. It was old property, with old-fashioned bay windows', only to be met by Mr Kidson's rejoinder: 'Artists come and paint that class of property, and photographers take photos'. In nearby Blackburn's Yard, Mr Bagshawe spoke eloquently about how conspicuous the site was in the townscape, in the centre of the view of the old town from the bridge, with 'buildings of very artistic interest', red-roofed and including the birthplace of Mary Linskill herself. T.H. Brown, speaking for threatened East Side owners, commented scathingly that 'the most searching inquiries should have been made to see what the East Side would be like if demolition orders were made and enforced ... before they destroyed Whitby's beauty, which was a great asset to the town. The general standard in Whitby was to be found in its narrow, steep yards, and he suggested that it was wrong to punish property owners because they owned such property'.[55]

This was unfair to Seaton Gray and Bewlay, whose concern to preserve as much as possible of Old Whitby was apparent, even if views at times had a higher priority than people. But Bewlay was also benefiting from rebuilding contracts, which raised potentially sensitive issues. Seaton Gray emphasised a dominant concern to avoid 'the too self-consciously picturesque' in replacement houses, but Councillor Jackson expressed widespread concern when he worried about the expense of the rebuilding programme, urging that it be done 'without any of the ancient buildings and fancy style' that he associated with Bewlay.[56] Resolving the contradiction was more difficult because the Ministry, while happy to pay lip service to preservation and to rebuilding in keeping with picturesque and historic surroundings, was unwilling to support the increased costs. Bewlay himself worried that new municipal housing on the East Side skyline, at the top of the cliff, would be too uniform (and too dominated by the emblematic red tiles).[57]

Such controversies helped to delay the first demolitions until 1938, in time to be interrupted by the Second World War, which (together with its austere aftermath) halted the process until the mid-1950s. Meanwhile, the Whitby fishing fleet had been quietly modernising during the inter-war years, using motorised cobles, with no complaint from the guardians of quaintness and the picturesque: the concerns of preservationists were entirely focused on buildings, layouts and vistas, and there was no attachment to traditional local fishing vessels, although the local coble accommodated itself well to the internal combustion engine. In a further irony, Dr Dale Wood announced in 1938 that Whitby's infant mortality rate was below the national average and that there had been no maternal mortality since 1930.[58]

When the war broke out, demolitions had begun, and other houses earmarked for disposal were vacant. Wartime housing shortages saw families moving back into the empty properties, while the cleared sites remained as ugly gaps in the urban landscape. In 1943, Seaton Gray looked forward to Whitby's post-war future as a fishing port and tourist resort, combining judicious improvement and the safeguarding of distinctive attributes and atmosphere. His 'Plan for Whitby' identified the 'essential requirements for the advancement of the town':

> First and foremost, must be the preservation of the unique atmosphere of the old-world Whitby, so that it doesn't become just one more concrete and plaster imitation of hundreds of other resorts ... a great many inhabitants, and ... members of the Council, do not realise that in its relics of the old time days of sailing ships, in some of its old buildings, even in its mountingblocks in the street, Whitby has something that *no other place has got*, and these things, so familiar to residents as to be regarded with a sort of affectionate contempt, are of the very essence of the place and of the very greatest interest and attraction to visitors.[59]

He was particularly exercised by the current problems of the East Side:

> This looks nearly as picturesque as ever it did, but behind its distant façade of beauty, there is a great deal of decay. A large number of houses are empty, having been condemned as unfit to live in – some 289 in all. There are many more which now show serious signs of decay....[60]

But Whitby had become 'by far the best known of the smaller resorts' because of advertising and exposure on the radio. There were plenty of picturesque and historic buildings to be preserved, including Grape Lane and Douglas Bolton's admired Tin Ghaut: 'I should like to see them both owned by the town. No alteration should be permitted to the few remaining old shop fronts, and when Grape Lane requires resurfacing, it should revert to cobbles or sets'.[61]

The post-war decade saw the completion of the demolition programme of the 1930s, and the construction of a few new houses on the East Side to Bewlay's designs. Some of these were so successful that, as McNeil later remarked, they deceived many onlookers into believing them to be authentic relics of Old Whitby.[62] But for more than a decade after the war, housing and materials shortages made further demolitions a practical impossibility, and complaints about the ugliness of the cleared or partially cleared sites rose to a crescendo in 1955.[63] At this point, the programme resumed, in keeping with central government policy for urban areas in general, as the cities were encouraged 'to start again on their big slum clearance programmes, which had been interrupted at the start of the war in 1939',[64] and new municipal housing became available to evicted tenants; the arguments of 20 years earlier were reprised, with some of the same *dramatis personae*, McNeil included. The Housing Act of 1954 gave clearance advocates a new target, identifying a further 400 'sub-standard' houses, as the frontier between acceptable and unacceptable was redefined in ever more demanding terms. The eventual fate of Grape Lane matched Seaton Gray's wishes, but Tin Ghaut was demolished and replaced by a car park, after a public inquiry which revisited all the issues of the 1930s. This followed closely after the demolition of a unique group of galleried houses on Boulby Bank (see Figure 8.3), which Bewlay had failed to recommend for preservation, but it was also the high tide of demolition. Whitby's Chamber of Commerce and tourist industry joined hands with the conservationists, who organised more effectively through the Literary and Philosophical Society and Whitby Preservation Society. The clearance and redevelopment party was driven back. Urban geographers recruited by the Marquess of Normanby, of nearby Mulgrave Castle, along with the Royal Fine Art Commission provided useful endorsement in the manner of the Bewlay Report but without explicit reference to it.[65]

The crucial issue, however, was the failure to sustain the building of new municipal housing in the early to mid-1960s, a hiatus which needs

Figure 8.3 The unique galleried houses of Boulby Bank, photographed by Frank Meadow Sutcliffe. These were among the casualties of post-war clearance and redevelopment.

further research. This bought time for the preservationists through an effective moratorium on further demolitions, while the post-war decline of the traditional holiday industry on the West Cliff placed greater weight on the kinds of tourists who were attracted to the harbour, as the balance between port and resort functions shifted in new ways. The fishing fleet continued to modernise, the herring fishery fell away after a boom in the mid-1950s, and pleasure boating grew in comparative importance. However, the idea of a fishing community by the harbour, and particularly on the East Side, was always more a matter of strategic visibility, romantic image and rhetoric than of hard demographic reality. By the end of the 1960s, as the pendulum began to swing nationally in favour of heritage and against modernist planning,[66] new markets emerged for housing in the harbour area, for holiday cottages and retirement homes, together with improvements in home improvement technologies, the easing of overcrowding (through smaller households and the knocking together of cottages) and the availability of improvement grants for reconditioning, which cumulatively transformed the prospects of Old Whitby. It had survived long enough to become viable in a new incarnation which sustained most of its picturesque appeal, while allowing its urban economy to revive in new but recognisable forms.[67]

Here is the key to Whitby's survival as a distinctive seaside resort with its own atmospheric sense of history and identity, which enabled it to be nominated as Britain's 'Best Seaside Resort' in 2006 by members of a well-established middle-class consumer information organisation.[68] Its current popularity with seekers of history, authenticity, atmosphere and the urban picturesque was not inevitable but has been the outcome of a series of conflicts and accidents, which resulted in the survival of enough of the old town to sustain a sense of romance and mystery among a broad spectrum of visitors, some more historically and architecturally informed than others. This has benefited from the survival of a local fishing industry, serving local outlets, alongside the development of pleasure boating in the harbour and a history of intermittent conflict between different harbour users, as well as a belatedly successful effort to remove sewage pollution from the harbour itself.[69] This chapter has delineated the processes by which the Whitby of the late twentieth century was able to overcome (or make a virtue of) its geographical isolation and economic and demographic stagnation, and take its place among a significant group of British seaside resorts which have defied prognostications of doom and decay and emerged as twenty-first-century success stories. Whitby is a remarkable illustration of the resilience and adaptability of the British seaside when it has a past to embrace, celebrate and exploit; and the part played by the fishing port in the renaissance of the resort, even as the remaining commercial traffic of the harbour declined into oblivion, was at the core of these developments. Whitby's story is not just an isolated case study but is an important example of a much wider and continuing trend.

Notes and References

1. E.W. Gilbert, *Brighton: Old Ocean's Bauble* (2nd edn., Hassocks: Flare Books, 1975), p. 95; F. Gray, 'Three views of Brighton as port and resort', this volume.
2. F. Gray, *Designing the Seaside* (London: Reaktion, 2007).
3. G. Shaw and A. Williams (eds.), *The Rise and Fall of British Coastal Resorts* (London: Cassell, 1997), especially chapters 4 and 5; John K. Walton, *The British Seaside: Holidays and Resorts in the Nineteenth Century* (Manchester University Press, 2000), pp. 163–5.
4. C. Beatty and S. Fothergill, *The Seaside Economy: the Final Report of the Seaside Towns Research Project* (Sheffield Hallam University, 2003), pp. 16, 23, 27, 51.
5. http://www.enjoyengland.com/ accessed 5 January 2007; http://www.whitbycottages.com accessed 5 January 2007.
6. Ian Carter, *Railways and Culture in Britain* (Manchester University Press, 2001).
7. Walton, *British Seaside*, pp. 1–3; Bill Bryson, *Notes from a Small Island* (London, 1995), pp. 124–5.
8. Colin Waters, *Whitby: a Pictorial History* (Chichester: Phillimore, 1992), Introduction; A. White, 'The Victorian development of Whitby as a seaside resort', *Local Historian* 28 (1998), pp. 78–93.
9. J.K. Walton, 'Fishing communities and redevelopment: the case of Whitby, 1930–1970', in David J. Starkey and Morten Hahn-Pedersen (eds.), *Bridging the North Sea: conflict and cooperation* (Esbjerg, Denmark: Fiskeri –og Sofartsmuseet, 2005), pp. 135–62; idem., *Tourism, fishing and redevelopment: post-war Whitby, 1945–1970*, University of Cambridge: Institute of Continuing Education, Wolfson Lectures, Occasional Paper No. 5, 2005; Peter Frank, *Yorkshire Fisherfolk* (Chichester: Phillimore, 2002).
10. Stephanie K. Jones, 'A Maritime History of the Port of Whitby, 1700–1914', Ph.D. thesis, University of London, 1982; Andrew White, *The Buildings of Georgian Whitby* (Keele: Ryburn Publishing, 1995).
11. Jan Hewitt, 'The "Haven" and the "Grisly Rokkes": Mary Linskill's dangerous landscapes and the making of Whitby', in T. Faulkner, H. Berry and J. Gregory (eds.), *Northern Landscapes: Representations and Realities of North-East England* (Woodbridge: Boydell, 2010), pp. 279–92; Mary Linskill, *In and About Whitby* (Whitby: Culva House, 2000).
12. M. McMillan, *Whitby Jet through the Years* (Whitby: Mabel McMillan, 1992); T. Woodwark, 'The rise and fall of the jet trade', *99th Annual Report of the Whitby Literary and Philosophical Society* (Whitby, 1921), pp. 25–34.
13. Stephen Bann, *The Inventions of History* (Manchester University Press, 1990), pp. 125–6.
14. Hewitt, 'The "Haven" and the "Grisly Rokkes"', p. 283.
15. A. and C. Black, *Picturesque Guide to Yorkshire* (Edinburgh: Adam and Charles Black, 1862), p. 355.
16. Ibid., pp. 353–9; *The Guide to Whitby and the Neighbourhood* (Whitby, 1850); Silvester Reed, *Reed's Illustrated Guide to Whitby and Visitors' Hand-Book* (Whitby: Silvester Reed, first edn. 1858); Martin Simpson, *A Guide to Whitby*

and the Vicinity (Whitby: Forth and Son, 2nd edn., 1887); Horne and Son, Limited, *Official Guide to Whitby* (Whitby: first edn., 1890).

17. *Seaside Watering Places* 1900–1, p. 34.
18. Frank, *Yorkshire Fisherfolk*, Chapter 8.
19. Clive Rouse, *The Old Towns of England* (London, 1936), p.89.
20. Horne's *Official Guide to Whitby* (Whitby: Horne and Son, 1934).
21. N. Pevsner, *Yorkshire: the North Riding* (Harmondsworth: Penguin, 1966), p. 388.
22. Walton, 'Fishing communities', pp. 143–5.
23. Pevsner, *North Yorkshire*, pp. 398–9.
24. Sue Berry, *Georgian Brighton* (Chichester: Phillimore, 2005); John K. Walton, *The English Seaside Resort: a Social History 1750–1914* (Leicester University Press, 1983), pp. 48–9.
25. R.H.J. Crook, *Walton-on-the-Naze, Essex: an Appreciation of a Picturesque, Old-World Seaside Resort* (London: Frederick Warne and Co., 3rd edn., 1911).
26. John Urry, *Consuming Places* (London: Sage, 1995); Bernard Deacon, 'Imagining the fishing: artists and fishermen in late nineteenth century Cornwall', *Rural History* 12 (2001), pp. 159–78.
27. Nina Lubbren, '"Toilers of the sea": fisherfolk and the geographies of tourism in England, 1880–1920', in D.P. Corbett, Y. Holt and F. Russell (eds.), *The Geographies of Englishness* (New Haven and London: Yale University Press, 2002); L. Newton, 'Cullercoats: an alternative north-eastern landscape?', in Faulkner *et al.* (eds.), *Northern Landscapes*, pp. 293–307.
28. B.E. Shaw (comp.), *Frank Meadow Sutcliffe, Hon. F.R.P.S.: Whitby and its People* (Whitby: Sutcliffe Gallery, 1974); M. Hiley, *Frank Sutcliffe: Photographer of Whitby* (London: Gordon Fraser Gallery, 1974); Peter Phillips, *The Staithes Group* (Nottingham: Phillips and Sons, 1993); Bulmer's *Directory* for the North Riding of Yorkshire (1890).
29. Charles Dellheim, *The Face of the Past: the Preservation of the Medieval Inheritance in Victorian England* (Cambridge University Press, 1982).
30. *Whitby Gazette*, 28 Feb. 1919; Brendan Rapple, 'James Russell Lowell and England', *Contemporary Review*, July 2000.
31. Frank J. Nash, *The Yorkshire Coast: its Advantages and Attractions* (York, 1904), pp. 11–12.
32. *Whitby Gazette*, 24 Dec. 1936.
33. Ibid., 13 Oct. 1933. See also Rouse, *Old Towns*, p. 89.
34. North Yorkshire Record Office, Northallerton (NYRO), MIC 1826, Bewlay Report and associated correspondence.
35. Jane Austen, *Sanditon* (London: J.M. Dent, 1978).
36. G. Douglas Bolton, *Yorkshire Revealed* (Edinburgh: Oliver and Boyd, 1955), pp. 217–23.
37. J. Delafons, *Politics and Preservation: a Policy History of the Built Heritage 1882–1996* (London: Spon, 1997), pp. 37–8, 62–4.
38. Quoted in *Whitby Gazette*, 24 Jan. 1947.
39. NYRO, Annual Reports of the Medical Officer of Health, Whitby Urban District Council, 1909 and 1914.
40. *Whitby Gazette*, 31 Jan. 1919.
41. *Whitby Gazette*, 21 Feb. 1919.
42. NYRO, Medical Officer's Report, 1930; National Archives, HLG 48/844, April 1928.
43. NYRO, Medical Officer's Report, 1929.
44. Walton, 'Fishing communities and redevelopment', pp. 146–9.

45. *Whitby Gazette*, 20 Oct. 1933, 27 Oct. 1933, 10 Nov. 1933.
46. *Whitby Gazette*, 3 Nov. 1933.
47. *Whitby Gazette*, 17 Nov. 1933.
48. *Whitby Gazette*, 19 Jan. 1934; Walton, 'Fishing communities and redevelopment', pp. 149–50.
49. *Whitby Gazette*, 25 March 1937; Walton, 'Fishing communities and redevelopment', pp. 150–1.
50. Scarborough Town Hall, Whitby UDC archive, Bewlay Report, August 1936.
51. Andrew Saint, 'How listing happened', in Michael Hunter (ed.), *Preserving the Past: The Rise of Heritage in Modern Britain*, (Stroud, 1996), pp. 115–33.
52. *Whitby Gazette*, 14 Aug. 1936; Walton., 'Fishing communities and redevelopment', pp. 151–2; NYRO, MIC 1826, Bewlay Report correspondence.
53. NYRO, MIC 1826, Bewlay Report correspondence.
54. *Whitby Gazette*, 27 Nov. 1936.
55. *Whitby Gazette*, 27 Aug. 1937, 15 July 1938.
56. *Whitby Gazette*, 5 March 1937, 25 March 1937.
57. NYRO, MIC 1826, Bewlay Report correspondence.
58. Frank, *Yorkshire Fisherfolk*, pp. 76–81; Ander Delgado and John K. Walton, 'La pesca y los pescadores en Inglaterra y el País Vasco, siglo XIX–1930: los casos de Whitby y Bermeo', *Itsas Memoria: Revista de Estudios Marítimos del País Vasco* 4 (2003), pp. 563–82; *Whitby Gazette*, 6 Nov. 1938.
59. V. Seaton Gray, 'Plan for Whitby', typescript dated 31 Aug. 1943, in Scarborough Town Hall, Whitby UDC Archives, pp. 8–9.
60. Ibid., pp. 2–3.
61. Ibid., pp. 8, 10.
62. *Whitby Gazette*, 12 April 1957.
63. *Whitby Gazette*, 18 March 1955.
64. Peter Hall, *Urban and Regional Planning*, (3rd edn., London, 1992), p. 125.
65. Walton, *Tourism, Fishing and Redevelopment*, pp. 15–24; G.J. Daysh (ed.), *A Survey of Whitby and the Surrounding Area* (Windsor, 1958).
66. P. Borsay, *The Image of Georgian Bath, 1700–2000* (Oxford: Oxford University Press, 2000).
67. Walton, *Tourism, Fishing and Redevelopment*, pp. 24–8.
68. M. Hilton, 'The organised consumer movement since 1945', in A. Chatriot, M.-E. Chessel and M. Hilton (eds.), *The Expert Consumer: Associations and Professionals in Consumer Society* (Aldershot: Ashgate, 2006).
69. Walton, 'Fishing communities and redevelopment', p. 142.

Chapter 9

Recycled Maritime Culture and Landscape: Various Aspects of the Adaptation of Nineteenth-Century Shipping and Fishing Industries to Twentieth-Century Tourism in Southern Norway

BERIT EIDE JOHNSEN

Introduction

This chapter focuses on the coastal region of Sørlandet (southern Norway).[1] Up to about 1880, Sørlandet was the centre for the Norwegian shipping industry. However, after this point, the industry declined: shipping and shipbuilding lost their importance as a direct result of the transition from sail to steamship, in particular the move from using wood to iron and steel as shipbuilding material. As a result, most traditional shipyards closed down during the 1890s, and the subsequent shipping crisis of the 1920s sealed the fate of the industry.

As one industry declined, another one – tourism – emerged. During the twentieth century, Sørlandet became Norway's most important summer seaside resort area, and both the coastal landscape and culture were 'adopted' by tourists. They 'invaded' the coastal areas, which included hundreds of tiny, rocky islands and skerries, remote beaches, bays and coves, small outports and seaside towns and cities consisting of charming, white lapboard houses.[2] They made new use and fresh interpretations of existing facilities: the ship owner's residence became a guesthouse; the captain's house became a summer house and fishing boats; and sailing ships were replaced by leisure boats in ports and harbours. The maritime culture of work was 'implanted' into leisure activities.

This chapter is divided into three parts. The initial focus is the cultural encounter between urban, bourgeois summer guests – also called bathing guests – and representatives of fishing and shipping communities in the decades on either side of 1900. Some relevant research on this cultural encounter between modernity (represented by visitors) and the 'traditional' and 'authentic' (represented by local residents) will be presented.[3] In particular, the contrast between production and consumption, work and leisure, as well as perspectives regarding year-round versus 'sun and summer' residence, will be highlighted.

The chapter's second focus is on how people's memories and reminiscences regarding traditional maritime culture were adapted to the tourism industry. What happened to the maritime cultures, land-scapes and townscapes after the traditional shipping industries had declined and after the modern ships had left their local shores and gone international? What happened after the fishing industry had entered into decline and changed dramatically during the twentieth century? What happened to traditional maritime culture when tourism continued to grow during the twentieth century? How was maritime culture – the immaterial as well as the material (including the maritime landscape and reminiscences regarding the shipping and fishing industries, architec-ture, maritime towns and outports) – used and reinterpreted by the tourism industry and, subsequently, tourists themselves?

In the last part of the chapter, the globalisation of maritime culture will be analysed, including tourists' particular perspective of the coastal landscape and townscape in a world of 'speed tourism' or 'mass tourism' emerging after the Second World War. Although this is a case study of Sørlandet, the aim of the chapter is to discuss general aspects as well as cast new light on the region's overall development from shipping and fishing to tourism.

Shipping Growth and Decline

From the eighteenth century onwards, and particularly during the nineteenth century, Sørlandet emerged as the centre of Norway's traditional shipping and shipbuilding industry.[4] For example, in the 1870s, Arendal (see Figure 9.1) was Norway's – if not Scandinavia's – largest sailing shipping port, operating with approximately 10% of the national tonnage. However, in the late 1870s, Norway – and particularly Sørlandet – was hit hard by a combination of stagnation and depression in the international freight market and a growing international trend towards using steamships. As sail was replaced by steam and wood by iron and steel across the world, the shipping industry of Sørlandet lost its competitive advantage and entered recession. The local yards closed down during the 1880s and 1890s, whilst several shipping companies either went bankrupt and closed down or moved out of the region. Surviving ship owners moved away from using locally built ships and bought foreign vessels instead. Ultimately, the shipping crisis of the 1920s sealed the fate of this traditional industry, and Sørlandet's ports saw a noticeable decline in their share of total Norwegian shipping

Figure 9.1 The port of Arendal *c.* 1870. The town was Norway's, if not Scandinavia's, largest sailing ship port at that time.

Figure 9.2 The crew of the four-masted iron barque *Songdal* (owned by the shipping company S.O. Stray & Co. 1912–1917) *c.* 1915. At this stage, Sørlandet still had a substantial fleet of **sailing** ships.

tonnage. In 1885, 33% of the Norwegian net tonnage was registered in Sørlandet (see Figure 9.2). By 1925, the region's ship owners managed a registration of only 5%. Nearly all the sailing ships were gone: They had been lost, wrecked or sold. Although the fishing industry was only of minor importance in southern Norway, the fishing crisis *c.* 1900 reinforced the same general tendencies.

In making the transition from sail to steam, the ship owners of Sørlandet were the slowest in the world to convert. However, from 1927 onwards, the region – in particular the largest towns – did experience a new tonnage growth based on motor and steam. Initially, this growth was linked to Anglo Saxon Petroleum Co., Ltd., and second-hand oil tankers. Although shipping never became as important as it once had been, this growth laid the foundation for a boom which lasted up to the 1970s. Nonetheless, modern shipping was fundamentally different from the traditional industry. Initially, most of the sailing ships had been locally built, manned and owned. Steamships and motor tankers, on the other hand, were owned by shipping companies located in the larger cities. They were normally built abroad, and they sailed worldwide. Although in the nineteenth century, sailing ships had been laid up for the winter in their home ports, the steam and motor ships seldom visited the local shores, and they no longer made use of the facilities found in the coastal yards and outports.

As the number of jobs fell dramatically, in particular from the 1880s onwards, thousands migrated not only to the United States but also to the Norwegian capital of Kristiania (Oslo) located in the eastern part of the country. During the second half of the nineteenth century, Sørlandet contained more than 8% of the Norwegian population, while in 1920 this proportion had shrunk to 6%. Yet within a few years, many people were returning as summer guests.

The rise of traditional shipping and related industries was primarily a pre-twentieth-century phenomenon, whilst the development of tourism was a feature of the twentieth century in Norway. Thus, tourism developed in coastal towns, cities and outports whose traditional economic roles were in decline. To a large extent, recreation and health became a substitute for a decaying shipping industry.

Cultural Meetings: Distance and Difference – The Scandinavian Research Tradition

The seaside holiday phenomenon originated in Britain and gradually spread to Scandinavia. Starting in the second half of the nineteenth century, representatives of the upper (and after a time, middle) classes in Scandinavian towns and cities (in particular, the capital cities) started to move out to the coast during the summer months.[5] The Danish and Swedish *bourgeoisie* started this tradition earlier than did the Norwegians. The resorts along the coasts of Bohuslän, Sweden, and Vendsyssel, Denmark, grew over a long period during the nineteenth century, eventually producing a breakthrough in the growth of boarding houses and summer hotels in the 1890s. A similar transition came three decades later (during the inter-war period) in southern Norway. Although some people visited the Norwegian coast for various health reasons during the late 19th century and a certain number of boarding houses were opened, representatives of the British upper classes (called 'salmon lords' because they were keen on fishing) as well as wealthy Norwegians preferred to scout out the rivers, deep fjords and high mountains of Norway during this same period.[6]

Behind the growth in seaside tourism lay a distinctive set of cultural, social, technological and economic developments: British and continental cultural influence, the introduction of summer holidays (the 'need' for vacations and paid holidays, particularly after the inter-war period), improved communication (better and cheaper transport: trains, cars, and regular coastal liner services) and improved economic conditions.

In Sweden, Denmark and (to a certain degree) Norway, highly interesting and valuable research has been undertaken on twentieth-century cultural encounters between local residents and visitors. The research contributions have different perspectives, including economic, political, social and cultural ones (although the latter is dominant). Researchers have analysed the encounter between representatives of different classes and ways of life or the encounter between *modernity* and *traditional forms of life*.[7]

It is my intention to focus in particular on the research carried out by Orvar Löfgren, Anders Gustavsson and Poul Holm. Based on their research contributions and articles on coastal communities of southern Norway, a distinctive, general pattern emerges.[8] Holidaymakers, called *bathing guests* or *summer guests*, were to be distinguished from tourists, who searched for the new and the unexpected whilst the former tended to favour familiarity, the replication of similar activities year after year and the reinforcement of group norms. Both groups came to well-established traditional maritime communities during the second part of the nineteenth century and the first part of the twentieth century.[9] Local people were occupied with fishing and/or shipping. The locals were therefore considered to be at *work* while the visitors were on *holiday*. *Work* contrasted with *leisure*, and visitors observed these *traditional activities*. At first, the local people and visitors regarded each other at a distance, registering social, economic and cultural *differences*. The attitude of the coastal dwellers towards the visitors was one not only of both *subservience* and *rejection* but also of *harmony* and *respect*. From the 1880s onwards, and particularly during the inter-war period (at which time the fishing and shipping industries suffered from a serious depression and went into decline), when the shipping industry finally went bankrupt, local residents became economically dependent on the visitors, although their economic situation did improve, particularly after the Second World War. Gradually, the locals' financial dependence on the urban dwellers decreased, transforming their inferior status into one of social, economic and cultural equality with that of the visitors. The result was a new consciousness among local residents, and these facts, along with the enormous pressure placed on the limited coastal resources (as a result of mass tourism and middle class purchases of summer residences), led to conflicts, particularly from the 1970s onwards.

The Cattegat–Skagerrak project (1981–1988), which brought together a large number of Norwegian, Swedish and Danish researchers from different fields of research, focused on the region's cultural development during the nineteenth century and the contact over the straits between Northern Denmark (Jutland, see Figure 9.3), Western Sweden (Bohuslän) and Southern Norway (Sørlandet).[10] The project published several reports on economic development, communication, shipping, fishing, migration and religious movements as well as tourism and cultural encounters between different social classes. The project's main focus was, however, on aspects of cultural development, or *cultural encounters*.

Project coordinator Poul Holm emphasised the 'discovery of the coast' by the *bourgeoisie* and stated that the bathing guests became the economic foundation for some coastal residents.[11] The success of the resorts

Figure 9.3 The fishing village Lønstrup, northern Jutland, Denmark, was gradually transformed into a summer resort from the late nineteenth century onwards. Fishermen and visitors shared the same beaches, a situation which might cause conflicts. Postcard from *c.* 1910.

depended upon being able to suppress rival functions such as fishing (or at the least make these activities look clean and romantic) in order that the resorts would not be seen as unclean workplaces. At the same time, the bathing guests complicated class structures and social relations between themselves and the newcomers as well as among the coastal residents themselves. Similar conflicts of interest – both commercial and social – are also discussed in Miskell's and Hussey's chapters in this volume. Holm compared the high level of conflict in the fishing communities of Bohuslän, Sweden, with the lower level of conflict in fishing communities found in northern Jutland, Denmark, making the comment that 'perhaps it has something to do with the coast not being "discovered" until a later date [around 1900], when the fishermen earned good incomes and therefore did not to the same degree feel urged to prostitute themselves for the urban guests'.[12] But this does not explain why the shipping communities of southern Norway also had fewer conflicts (at least up to 1970). An alternative explanation for this absence is outlined below.

In Sørlandet, the growth of tourism from the 1920s onwards coincided with the closing of the shipping industries. Thus, there was no need to suppress or kill off rival functions, because the maritime villages and outports were already nearly emptied of both traditional maritime activities and traditional maritime labour. They became natural and clean all by themselves, without the 'help' of the holidaymakers – a type of help which might have caused conflicts. The coastal communities were ready to be invaded by new groups, ready to be recycled according to different needs and a new culture – the *bourgeoisie*'s culture of holidaymaking (see Figure 9.4). At the same time, the region desperately needed alternative income opportunities. Thus, shipping communities in Sørlandet adapted to the newly expanding tourism industry with less conflict of interest than did regions in which maritime industries such as fishing still existed, making use of the same shores (islands, ports, harbours, outports, etc).

There were – and are – differences between regions where the shipping industry declined, closed down or left the coast and went international in tourism's infancy (or at least before tourism established itself as a major industry), similar to Sørlandet, and regions where maritime industries (particularly fishing) developed and are still simultaneously developing alongside tourism (e.g. northern Jutland and Bohuslän). In several regions all over the Western world where maritime industry is only history, reminiscences of the maritime past have been easily adapted, reinterpreted and changed according to others' needs and standards. In areas where the fishing industry developed in conjunction with an expanding tourism industry, selling much of its product to this market, there was a different – difficult and often ambivalent – situation, resulting in not only

Figure 9.4 One of Sørlandet's earliest boarding houses, Fevig Sommerhjem (Fevig Summer Home), was opened in 1889 in a former ship owner's residence at Fevik near Arendal. In 1895, the name was changed to Fevig Kystsanatorium (Coastal Sanatorium) and later Fevig Bad (Bath). The boarding house attracted well-off middle class people from the larger towns of Norway.

internal conflict but also interdependence. Production was dependent on consumption; the fishing industry was dependent on the ever-expanding tourism industry. The difficulties arose from the fact that tourism normally was the leader – the stronger partner – and the fishing industry therefore became more and more dependent on an unreliable 'friend'.

As shown, several Scandinavian studies on tourism have analysed the encounter between visitors and locals as a *cultural encounter* and have only to a certain extent included social and economic aspects in their analyses. They have not focused on distinctiveness and contact between two existing groups of people and two forms of life but on how the newcomers – the tourism industry and holidaymakers – invaded the original, authentic culture and adapted it to their own requirements and images. Moreover, they have not analysed what happened to the reminiscences of the maritime culture *after* the maritime industries had fallen into decline. In accordance with an outline of tourism development in southern Norway from 1900 up until today, the (re)invention of maritime culture in a tourism setting during the 20th century will be analysed in more detail.

The (Re)invention of the Maritime Tradition in a Tourism Setting

The development of southern Norway's tourism industry from the late nineteenth century until today and the uses of maritime images as collective memory and heritage may be divided into four periods:

(1) 1880–1910: Introduction of coastal tourism.
(2) 1910–1930: Discovery of Sørlandet as a tourist destination and the introduction of maritime images as heritage.
(3) 1930–1990: 'Sun, Summer, Sørlandet' – Development of romanticised images and entry into a collective memory.
(4) 1990–today: New definition of Sørlandet, adjustment of (but also a struggle against) traditional images.

Along with the urban *bourgeoisie*, artists (among them several famous writers and painters) were pioneers and trendsetters when they were summer guests during the early twentieth century.[13] This coincided with the introduction of the regional name 'Sørlandet' (The Southern Land), which was introduced by author Vilhelm Krag (1871–1933) in 1902.[14] Up to that point, the region had simply been called 'Agder' and was regarded as being a part of western Norway. However, Krag's ambition was to create a strong southern Norwegian identity, among other things, by establishing regional institutions such as a museum, an archive and a historical association. In doing so, Sørlandet would be firmly anchored in its proud past. Furthermore, it was no coincidence that he introduced his great project shortly before the union between Norway and Sweden was dissolved in 1905. A strong sense of nationalism influenced Krag's (and others') regionalism. In Norway, regionalism (the periphery as opposed to the centre) was an important part of the country's political struggle throughout the 20th century (and to a certain extent still is).

In the decades following the introduction of the region Sørlandet in 1902, a common history and identity had to be created or 'invented'.[15] Simultaneously, tourism grew, and as a consequence of this growth, both local inhabitants and visitors took an active part in the growth process. The proud maritime past was crucial to this development, constituting important elements in the creation of this identity. However, the particular elements of the identity – the various components of the maritime past which were highlighted, recycled or (re)invented – changed through the years.

As already mentioned, the traditional nineteenth-century shipbuilding and shipping industries (in contrast to the modern steam and motor ships from the 1930s onwards) had made extensive use of the coastal landscape. During the inter-war period, the vitality of the traditional maritime economy and culture in southern Norway had vanished, and at this point, the reminiscences of the old shipping industries were ready to be stored in both the Norwegian Maritime Museum and other such museum collections. Moreover, while archives and historical societies (many of them established during the first part of the twentieth century) were prepared to preserve these memories, there existed several other optional and alternative uses of them as well.

Tourism expanded during the second time period, particularly during the inter-war years, as both the sea and sun became attractive. The health aspect was gradually replaced by one focused on relaxation, fun and pleasure (as opposed to work). Many hotels, small boarding houses and private guest houses with basic standards were established along the coast of Sørlandet.[16] In fact, the seaside boarding house offering modest accommodation became a Nordic speciality.[17]

Reisetrafikkforeningen for Sørlandet (Southern Norwegian Travel Association) was established in 1928. One year later, Sørlandet had 900 foreign visitors; in 1935, it had 2000 visitors; and in 1939, 5241 foreign guests visited the region, with the majority being Danes and Germans.[18] These figures were probably accompanied by a considerable growth in the number of Norwegian visitors (who were the overall majority of summer guests), although exact figures do not exist. Most visitors came by coastal steamers from Eastern Norway, but some also came by ferry from abroad. Finally, in 1938, the railway line between Oslo and Kristiansand opened.[19]

As in other coastal settings, summer guests provided a valuable source of extra income for local residents. Local people rented out rooms in their private houses. They acted as servants, catering for well-off middle class people from the larger cities, particularly from Oslo, Norway's capital. Furthermore, local people worked in their everyday environment while holidaymakers relaxed away from home, which influenced their tourist-oriented 'gaze'.[20] As a result, the name 'Sørlandet' became synonymous with going on holiday to enjoy the sun and to experience tranquillity.

Next, even though the sailing ships were history after the First World War, they still existed in the dream world of tourists as well as that of many local residents. In songs, poems, fictional stories and non-fictional accounts, the images of the white sails' heyday lived on. In addition, small towns, villages and outports – marked by their characteristic nineteenth-century architecture (mostly white but also red and yellow lapboard houses with small, square-paned windows) – reminded visitors of the region's proud maritime past. The fact that Vilhelm Krag published a book called *De skinnende hvide Seil (The Shining White Sails)* in 1931 was no coincidence. The title of an article in a regional magazine in 1930 also caught the essence: 'Take Care of the Memories' – the memories of the proud maritime past.[21] Tourism – and the image of

Sørlandet as a holiday paradise – lived on and expanded, and the past was revised in accordance with a changing present. The maritime past was reinvented – and recycled – in a continuous process.

Tourism continued to grow after the Second World War, during the third period and particularly from the 1960s onwards as a result of better communication, general economic growth and longer holidays for ordinary working people as well as new leisure trends.[22] Most employees were granted two weeks' paid holiday during the 1930s, and all employees were allowed to take three weeks of holiday starting in 1947. Simultaneously, the growth of new industries, improved external transport and overall improved economic conditions along with social adjustment changed the uneven balance between holidaymakers and local residents. Unemployment and poverty caused by the slump in maritime industries no longer forced the local population to serve well-off middle class tourists and summer guests. The days of inferiority and dependency were over. Accordingly, a new, higher level of self-esteem emerged among local people.

The growth in Mediterranean holiday packages from the 1960s onwards did not cause a decline in domestic holidays; rather, the contrary occurred. However, charter tourism did change holidaymaking patterns. In 1964, Norwegian workers were granted four weeks of holiday. Furthermore, the general Norwegian standard of living had reached a considerably higher level than previously. The traditional boarding houses could no longer meet the market's demand for higher standards of accommodation. They soon became old-fashioned and outdated, and it did not always pay to renovate them, so many closed down. Instead, many Norwegians – some of them former boarding house guests, particularly from eastern Norway – bought or built their own summer houses. In addition, modern hotels were built while other older ones were renovated. The number of campgrounds also grew.[23] In addition, the increasing number of privately owned cars highly influenced tourism. In 1950, there were only 65,028 private cars registered in Norway, while in 1960, there were 225,439; and in 1970, that number had risen to 747,966.[24]

Today, well into the fourth period, the popularity of the coast of Sørlandet within Norway, not least among persons living in Oslo, makes it a holiday destination that features on the front pages of national as well as regional and local newspapers, particularly during the spring and summer months. During recent decades, there has been a contest over the use of space, particularly in those areas in close proximity to water. Pressure upon the coastline, rising prices of modest summer houses (often called skipper houses, although several have never been captains' homes at all) as well as luxurious residences by the sea have been repeated discussion topics. From 1965 onwards, there has been a general legal ban on building within the 100-metre coastal belt. Likewise, it is difficult to obtain permission to build new summer houses, although dispensations are occasionally granted. Another topic attracting considerable discussion concerns the recently introduced local residence obligation (*boplikten*, a duty to live permanently in year-round houses) which has put severe restrictions on the right to use traditional houses as summer houses. Consequently, the price of this scarce commodity has soared.

To some extent, distinctive social – and socio-economic – groups are patronising different kinds of resort areas, and there is also internal social zoning within these areas. Nevertheless, Sørlandet and its inhabitants have traditionally been characterised by social and economic equality. To a certain extent, *Allemannsretten* (The right of public access) makes the coastline accessible to everybody, subject to good behaviour and respect for property and privacy.[25] But this ethos of equality has been challenged during recent decades by the *nouveaux riches* – permanent residents as well as holidaymakers.

On another front, more than one million people take the ferry between Hirtshals (Denmark) and Kristiansand on an annual basis.[26] In June, July and August, Sørlandet offers more than one million people accommodation at hotels, campgrounds or cottages, 85% of whom are Norwegian (compared with 70% on a national basis).[27] However, only a small percentage of the holidaymakers are officially registered since thousands of summer guests and tourists visit Sørlandet and stay in the private homes of either their family or friends (or in their own summer houses). Sørlandet covers less than 5% of Norway but has more than 8% of the country's leisure homes. The coastline has the largest density of summer houses compared with the hinterland, and visitors outnumber the permanent residents during the summer months. In fact, today's holidaymakers also include permanent residents themselves. It is simply both acceptable and popular to spend the summer holidays at home or close to home. These numbers give an idea of the importance of tourism in southern Norway.

While Norwegians took 5.8 million holiday trips in 2007 (about 600,000 travellers went to Spain alone),[28] they also favour their own coast and maritime culture. Many summer guests have been regular visitors to southern Norway for many years, and they return every summer. It is a journey back to the world of their forefathers; it is childhood and family history. They have a tendency to construct a romanticised past, and for many people, the maritime past of their forefathers and the 'days of the white sails' form an essential part of this construction.

Invention of the Maritime Tradition in a Tourism Setting

How was maritime culture, history and landscape reinterpreted and recycled during the twentieth century? As shown, the maritime culture of Sørlandet gradually ceased to be a dynamic force within active shipping, shipbuilding and fishing industries after 1900. With reference to Eric Hobsbawm's discussion of 'the invention of tradition', it was freed of its fully symbolic and ritual use when no longer fettered by practical use.[29] The maritime tradition was ready to be reinvented.

During the inter-war period, and gradually after the Second World War, certain images of Sørlandet developed that were particularly influenced by the region's popularity as a holiday destination. The most popular images found in this third period were romanticised ones: 'Sun, Summer, Sørlandet'. Even today, the most frequently marketed images of Sørlandet are those containing 'humble and slow-moving Southerners', 'holiday, sun and summer', 'nostalgia and romance' and 'nineteenth-century shipping, fishing and architecture'.[30] Through this form of marketing, 'Sørlandet' gradually became a trademark, a tourism product.

Generally speaking, traditional maritime references, symbols and stereotypes have been essential to the general image-building of Sørlandet, and they still are. They have frequently been used in the marketing of different local products and destinations. The maritime rituals have, to a large extent, remained static, and the artefacts, landscapes and townscapes of Sørlandet have survived more or less unaltered throughout the twentieth century. But their meaning and interpretation have changed profoundly over the years, depending on the nature of the context, and I intend to illustrate this point by citing a few examples taken from recent events.

Firstly, the idea of labelling Lillesand as one of the best-preserved 19th-century maritime villages, 'The Sailing Ship Village' or 'Slow City' (see below), has undoubtedly certain elements of maritime nostalgia. This small village has, among other things, a sailors' choir (*Hermanos*); however, the majority of these choir members have never been sailors at all. Nevertheless, they wear sailor's uniforms, meeting the audience's expectations of a genuine maritime song tradition. The choir members not only sing traditional folk songs but also supplement their repertoire with new songs taken from the same genre.[31] They perform 'invented traditions', and they have adapted to modern times.

Secondly, there are only three large sailing ships left in Norway; these are *Christian Radich* (Oslo), *Statsraad Lehmkuhl* (Bergen) and *Sørlandet*, all with steel hulls. Thousands of wooden sailing ships were built during the 19th century, particularly along the coast of southern Norway, and yet not a single one has survived. *Sørlandet*, the sailing ship representing southern Norway, was built as a training (school) ship in 1927 and remained as such until 1974. She never sailed overseas in commercial trade during the sailing age. But for all the people who gaze upon her in the harbour of Kristiansand, sail on chartered tourist voyages or join overseas cruises during the summer season, she represents the proud maritime past of Sørlandet – the sailing ship era.[32] Thus, another suitable tradition has been 'invented'.

Similarly, a number of small ferry boats were lifelines between the mainland and the small outports and islands along the coast of Sørlandet until the 1980s and carried mail and goods as well as passengers all year around. However, during the 1970s and 1980s, the number of regular inhabitants on the islands was drastically reduced, and these passengers were gradually replaced by tourists. Consequently, work and everyday functions were replaced by leisure and holiday pursuits. Today, some of these small boats still sail during the summer season, representing traditional coastal culture and thereby 'invented tradition'.

A further example concerns lighthouses, which for centuries were essential for ships' security when travelling along the dangerous Norwegian coast. During recent decades, as modern maritime technology has made these lighthouses more or less redundant (and they are no longer inhabited), they have been recycled into nostalgic boarding houses, representing not so much security as they do nostalgia and 'green' tourism.

Initially, this question was raised: how *were* – or perhaps more important, how *are* – maritime culture, history and landscape recycled, reinterpreted or reinvented? John Urry argues that the answer to this question lies in the present, as tourists see – or gaze at – objects constructed as signs. The signs 'stand for something else'.[33] The picturesque towns and outports, the full-rigged sailing ships, the sailors' choirs, the ferry boats and the lighthouses have something in common; they are all signs of Sørlandet's proud maritime past and are interpreted as representing the area's traditions and emphasising its continuities – our collective memory and heritage. The maritime past adapted throughout the twentieth century to ever-changing social, cultural and economic conditions. This adaptation was influenced by both national and global forces, which leads me to my third and final point.

Post-Modern Tourism and Global Stereotypes

After having visited Sørlandet in June 2002, journalist Siobhan Mulholland wrote an article that appeared in the London newspaper *The Independent*.[34] Mulholland found that Norway conveys to most of its 125,000 annual British visitors the ideas of skiing, snow-covered mountains, western fjord cruises and traditional knitwear. But in Sørlandet, she found 'one of Scandinavia's best kept secrets'. She emphasised the typical and well-preserved architecture as well as the geography – no huge hotels or high-rise apartment blocks but rather summer houses built in the local style. 'Why do so few foreigners visit this area?' she asked. She found that the cold weather and high price levels were the main reasons why the area attracted so few foreign visitors. The exclusiveness of the region is why she described it as 'the Norwegian Riviera', with flashy yachts, wealthy visitors and exclusive residences.

To cite another example: in August 2002, *New York Times* critic Anthony Tommasini visited *The Risør Festival of Chamber Music*. In December, he ranked the festival as number two among the world's most memorable 'Moments in Classical Music', not only because of the music itself but also because of the 'idyllically located fishing village'.[35] But this was a misapprehension since, with the exception of a certain amount of inshore fishing taking place in the Risør area today, this small town has actually never been a fishing village. Risør prospered during the nineteenth century because of its shipping, shipbuilding and timber industries, just like all the other small towns and villages along the Skagerrak coast.

An important question to ask regarding this topic is whether and to what extent the towns and outports of southern Norway should adapt to global stereotypes such as 'fishing village', 'picturesque maritime community' or 'picturesque shipping town'? The 'picturesque' aspect is what often strikes a non-Norwegian visitor. Norwegians also find their ideal summer dream in the traditional towns and outports characterised by nineteenth-century architecture. Apparently, there is great market potential here, and an adaptation of the towns, villages and outports as 'nineteenth-century theme resorts' could be done quite easily, since they are relatively (some would say very) well preserved. For example, the small outport of Lyngør, situated on four islands in the eastern part of Sørlandet, was awarded a prestigious prize as 'Europe's best kept village' in 1991.[36]

Growth and development was so modest and relatively unobtrusive during the twentieth century that many towns, villages and outports still stand in close comparison with their nineteenth-century photographs. Three concepts involving Lillesand (see Figures 9.5 and 9.6) have been suggested: 'Slow City', 'The Sailing Ship Town of Lillesand' and 'The Picturesque Towns of Skagerrak' (with reference to the white painted towns, often called 'a string of pearl necklaces'). The 'Slow City' movement was founded in Orvieto, Italy, in 1999, and it has grown into an international network including some 32 cities.[37] The concept is based on values such as quality of life, local identity and roots, traditional food, culture and architecture – these in contrast to global trends, consumer mentality and mass tourism. For example, neither neon lights nor motor vehicles are allowed within city limits. In November 2002, some local citizens suggested that Lillesand should become part of this movement, thus becoming a part of a global maritime heritage dream world.

Local traditionalists tend to favour a repackaging or adjustment of Sørlandet's traditional images: 'humble Southerners', 'holiday, sun and summer' and 'nostalgia and romanticism'. They want to build sites of collective memory and heritage through focusing on nineteenth-century shipping and architecture. However, this type of retrospectively based

Figure 9.5 The small village of Lillesand, 1909. The sailing ships and maritime activities mark the port.

Figure 9.6 Lillesand, 2009, still a small village. Leisure boats have substituted the sailing ships, but the village has been relatively well preserved, and it has an authentic nineteenth-century look.

development has met some resistance, with certain people even labelling the 'Slow City' concept as being reactionary. Referring to the long, cold winter and the lack of activity, they state that 'we're already slow enough – we should speed up instead!'[38] Businesspeople and academics alike want the region to be related to international business and industry and a modern university, among other things, as well as a future-oriented perspective.[39] They challenge the images created by traditional writers such as Vilhelm Krag and preserved by the *nouveaux riches* from eastern Norway, particularly the Oslo region. They want to link present needs and future visions to prosperous and industrious images – growth and work, not leisure and relaxation. Accordingly, not everyone would like the villages and outports of Sørlandet to be repackaged as romantic, historical and authentic tourist destinations or theme resorts.

Conclusion

In this chapter my primary focus has been the cultural encounter between the urban *bourgeoisie* as summer guests and representatives of fishing and shipping communities during the decades surrounding 1900. My second and third points of focus have included certain aspects of the development and interpretation of maritime culture later on in the twentieth century, when the traditional shipping industries were in decline and when tourism continued to grow. Today, Sørlandet is the most important summer holiday resort area for Norwegians. While tourists and tourism positively affect the local scene, they simultaneously apply a great deal of pressure to it.

Why has the maritime past been particularly attractive when traditions were invented during the twentieth century in Sørlandet? One aspect lies in the fact that local people and newcomers as well as tourists and summer guests occupied the coastal landscapes and townscapes already characterised by their maritime activities, finding reminders of the past there. However, an additional, important aspect is the fact that maritime history and culture has been associated with wealth, openness and international contacts, attributes that are considered attractive in terms of both values and ways of life.

Finally, in my discussion of present trends, I have particularly focused on both the globalisation of tourism and the development of maritime stereotypes. Moreover, the interpretation and use of the southern Norwegian coastal landscape; reminiscences regarding shipping and fishing industries; architecture of towns, ports and outports as well as musical and dance traditions are all under the influence of globalisation. The present develops in a continuous process influenced by the past. This process will continue to be influenced by tourism, currently the world's largest and fastest growing industry.

Notes and References

1. In January, 2008, the total population of Sørlandet (the two Agder counties) was 272,000. Kristiansand, the largest city, had 78,919 inhabitants. Statistisk Sentralbyrå (Statistics Norway), http://www.ssb.no accessed 15 November 2008.

2. The small towns/cities and villages along the coast are from east to west: Risør, Tvedestrand, Arendal, Grimstad, Lillesand, Kristiansand, Mandal, Farsund and Flekkefjord.

3. For a discussion of authenticity in relation to tourism, see D. MacCannell, *The Tourist. A New Theory of the Leisure Class.* (New York: Schocken Books, 1976); E. Cohen 'Authenticity and commoditization in tourism,' *Annals of Tourism Research* 15 (1988), pp 371–386.

4. This chapter is based on Berit Eide Johnsen, *Rederistrategi i endringstid. Sørlandsk skipsfart fra seil til damp og motor, fra tre til jern og stål. 1875–1925. (Shipowners' Strategies in a Time of Change. The Shipping Industry of Sørlandet, Southern Norway, from Sail to Steam and Motor, from Wood to Iron and Steel. 1875–1925.)* Kristiansand: HøyskoleForlaget (Norwegian Academic Press, 2001).

5. A. Brodie and G. Winter, *England's Seaside Resorts* (Swindon: English Heritage, 2007); P. Holm, *Kystfolk. Kontakter og sammenhænge over Kattegat og Skagerrak ca. 1550–1914.* (With an English summary). (Esbjerg: Fiskeri- og Søfartsmuseet. Saltvandsakvariet, 1991), p. 308.

6. One of Sørlandet's earliest boarding houses, Fevig Sommerhjem [Fevig Summer Home], was opened in 1889 in a former ship owner's residence at Fevik near Grimstad. In 1895 the name was changed to Fevig Kystsanatorium [Coastal Sanatorium]. H. Austarheim, (ed.) (2002) *Sol, Sommer, Sabeltann. Sørlandsturismen i 100.* (Arendal. 2002).

7. There is, for example, an extensive amount of literature dealing with Danish fishing villages and their transition to tourist destinations during the last part of the nineteenth and first part of the twentieth century. Most of this literature has an ethnological perspective. The 'cultural meeting' between locals and visitors – the former fishermen and their families, the latter summer guests or bathing guests – is analysed, but sometimes only very briefly, as in K.M. Hansen, *Tilbage til turismens rødder. 150 års badeturisme langs den nordjyske vestkyst.* Nordjyllands amt, 2002); J. Slettebo (ed.) (1985) *Sommerglæder. Dansk Kulturhistorisk Museumsforening hylder Peter Seeberg på 60 års dagen og tilegner ham bogen..*(Sønderborg/Viborg, 1985).

8. A. Gustavsson, P. Holm and H. Try (eds.) *Kulturmøtet mellom badegjester og fastbuande / Kulturmötet mellan badgäster och fastboende. Kattegat-Skagerrak Projektet Meddelelser.* (No. 3. Aalborg 1983. No.5. Aalborg 1984. No. 13, Kristiansand 1987); Holm 1991; O. Löfgren, *Fångstmän i industrisamhället. En halländsk kustbygds omvandling 1800–1970* (Lund, 1977); O. Löfgren *Fiskarna vid Båtfjorden.* (With an English summary) (Varberg: Varbergs Museum, årsbok, 1969); A. Gustavsson, *Sommargäster och bofasta.* Stockholm, 1981); K. Grindland, (pp. 43–67) '... at raade Bod Livskraften'. *Agder Historielags Årsskrift* (Kristiansand, 1990); Slettebo (1985), and B.E. Johnsen,'Med lua i handa? Ferierende og fastboende gjennom hundre år.' in J.P. Knudsen and H. Skjeie (eds.): *Hvitt stakitt og fiberoptikk. Regionale myter – regional makt.* Kristiansand: HøyskoleForlaget (Norwegian Academic Press, 2002), pp. 249–272.

9. Gustavsson (1981).

10. The full name of the project was 'The Cultural History of the Coasts along the Cattegat and the Skagerrak in the Nineteenth Century'. The perspective was contact over the seas, Skagerrak and Cattegat, between Northern Denmark (Jutland), Western Sweden (Bohuslän) and Southern Norway (Sørlandet), originally inspired by Henri Lefebvre's theories.
11. Holm (1991), pp. 275–294 and 308 (English summary).
12. Holm (1991), p. 308.
13. From 1900 onwards, but particularly during the inter-war period, artists' colonies were established along the Skagerrak Coast, in the small town of Mandal and the outports Flosta (near Tvedestrand), Narestø and Tromøy (near Arendal), Rønnes (near Grimstad) and Brekkestø (near Lillesand), among others.
14. Krag published an article in the Kristiania (Oslo) newspaper *Morgenbladet* 16 March 1902, titled 'Nordmænd' ['Norwegians']. *Sørlandet* covers the two counties Aust-Agder and Vest-Agder, both the Skagerrak coast and the hinterland. Before 1902, Agder had been considered to be a part of western Norway - *Vestlandet*. Krag argued that Sørlandet had its own identity and history, different from western Norway. See J. Andreassen, *Sørlandet og Vilhelm Krag. Tekster* (Kristiansand: Vilhelm Krag-selskabet, 1996), p. 17.
15. E.J. Hobsbawm and T. Ranger (eds.) *The Invention of Tradition.* Cambridge: Cambridge University Press, 1989 ed.), pp. 1, 4, 263. For a discussion of Hobsbawm's term 'invention of tradition' with reference to Sørlandet, see B.E. Johnsen, 'Images of Sørlandet: The Skagerrak Coast of Norway. Representations and stereotypes, and the fight against them,' in A.Kostiainen and T. Syrjämaa (eds). *Touring the Past. Uses of History in Tourism.* Discussion and Working Papers No 6 (The Finnish University Network for Tourism Studies (FUNTS). Savonlinna. 2008).
16. Flekkerø Skjærgaardssanatorium was established in 1903 near Kristiansand as one of Sørlandet's first summer sanatoriums (not a hospital, but a boarding house where well-to-do town dwellers could recuperate), and was followed by several other sanatoriums, boarding houses and hotels. See Austarheim (ed.) (2002), pp. 15–21.
17. O. Löfgren, *On Holiday. A History of Vacationing* (Berkeley: University of California Press, 1999), p. 121.
18. A. Haddeland, (ed.), *Ekstrakt av Reisetrafikkforeningen for Sørlandets arbeide gjennom 12 år - 1928–1940.* [The Travel Association for Sørlandet]. (Kristiansand, 1940).Haddeland 1940: 12, 19.
19. The railway between Oslo and Kristiansand, *Sørlandsbanen*, had been under construction for decades. The line from Oslo to Arendal opened in 1935. Kjevik Airport opened in 1939. There were also ferry connections between Kristiansand and Denmark (Hirtshals) and England (Newcastle) in the inter-war period. See Austarheim (ed.) (2002), p. 8.
20. J. Urry, *The Tourist Gaze* (London: Sage, 2002 edn.), p. 1.
21. Johnsen (2008), referring to Lindeberg 1930: 'Vern om Minderne'.
22. Austarheim (ed.) (2002), p. 8.
23. The number of campsites grew, particularly from the 1960s onwards, when cars became common, together with tents and later caravans. Austarheim (ed.) 2002.
24. In 1999, the number of cars was 1,813,642. Motor vehicles 1950–1999. Statistisk Sentralbyrå (Statistics Norway), http://www.ssb.no Accessed 15 November 2008.

25. The Norwegian Outdoor Recreation Act of 1957 legalised the right of access to the public.
26. 14% of the 1.1 million passengers between Hirtshals (Denmark) and Kristiansand (Norway) were Danish, 18% German, and 50% of the traffic took place in June, July and August 2000. Color Line statistics.
27. Fædrelandsvennen 15 October 2003. In 2002, Sørlandet for the first time had more than two million commercial guest nights (the whole year). Fædrelandsvennen 5 March 2003 with reference to Statistisk Sentralbyrå (Statistics Norway) 2002. In July 2002, 93% of the tourists in Kristiansand were Norwegians. Statistics from Destinasjon Sørlandet, Kristiansand.
28. Statistisk Sentralbyrå (Statistics Norway), http://www.ssb.no/emner/10/11/reiseliv/main.shtml Accessed 13 October 2008.
29. Hobsbawm and Ranger (1989), p. 4.
30. For a further discussion of the (tourism) images of Sørlandet, see Johnsen (2008).
31. See Hobshawm and Ranger (1989), p. 6. He mentions different traditional practices, among them folksongs, which were 'modified, ritualized and institutionalized for new national purposes'.
32. The idealistic association 'Fullriggeren Sørlandet' has owned the full rigged ship from 1981.
33. Urry (2002), pp. 12–13 and 117: For example, a 'pretty English village can be read as representing the continuities and traditions of England from the Middle Ages to the present day.'
34. *The Independent* 27 June 2002. Mulholland reffered to Sørlandet as 'the Skagerrak coast of Southern Norway'. See also Urry (2002), p. 86: 'A reflection of this attraction of the real or natural in tourism has been the 'Campaign for Real Holidays' conducted in one of the key newspapers of the British service class, *The Independent*.'
35. *New York Times*, December 29, 2002.
36. Lyngør grew during the nineteenth century and several hundred people inhabited the outport. Today, only 90 people inhabit Lyngør all year round. During the summer months, however, this number increases to several hundred. *The Independent's* journalist (see earlier note) called this outport one of the world's best holiday destinations.
37. The Slow City movement (originally Italian, Cittaslow) includes 32 cities situated in Croatia, Germany and Norway (Levanger) apart from Italy. *Fædrelandsvennen* 5 November 2002.
38. *Fædrelandsvennen* 7 November 2002.
39. Lobbyists from the region succeeded in establishing the University of Agder (formerly Agder University College) in 2007. Successful lobbying was also the main reason why a new road connecting the southern coast from east to west is currently under construction (2007–2010). The Sørlandet region has an important cluster of internationally oriented offshore and drilling engineering businesses as well as an extensive oil, gas and processing industry, among other things.

Chapter 10

Gijón: From Asturian Regional Port and Industrial City to Touristic and Cultural Centre for the European Atlantic Arc, from the Nineteenth Century to the Present

GUY SAUPIN

Gijón, with 270,000 inhabitants, is the most important city of Asturias, an autonomous region in northern Spain, and its demographic growth surpasses that of Oviedo (200,000 inhabitants), the political capital of the Principality, and leaves that of Avilés, the other great industrial port, far behind. These three points of the region's central triangle contain, if we include their surrounding districts, three-quarters of the population of Asturias, which amounts to around a million in total. In contrast with that part of the Principality where the rural economy survives in significant strength, Gijón, like the other cities of the central zone known as 'Ciudad Astur', presents an employment structure in conformity with the average for the European Community before its most recent expansion (see Table 10.1).[1] Gijón pulls together around a quarter of the regional economy, but the crisis it has experienced in its basic industries has left it lagging behind Oviedo in per capita income.

Gijón is a port city which developed above all from the nineteenth century for the export of coal from the mines of Asturias, whose rich energy resources formed the basis for an initial phase of industrialisation which pulled together metalworking industries, shipbuilding, textiles, glass and foodstuffs. The decline of the Asturian coal industry was compensated by imported fuels, including coal of more suitable quality, which sustained a second, very strong phase of industrialisation based on new technologies for steel manufacture, particularly during the 1960s and 1970s. The industrial crises which have struck the steel, shipbuilding and textile industries have had a terrible impact on the city's economy and society, requiring local leaders to engage in extensive reconsideration to develop new perspectives and allow Gijón to bounce back.[2] In this redefinition of the motive forces of future development, the maritime character of the city and the rise of

Table 10.1 Employment by sector

	Agriculture and fishing	Manufacturing and construction	Services
Gijón	1.9%	32.1%	66.0%
'Ciudad Astur'	2.9%	30.8%	66.4%
Rest of Asturias	38.1%	20.7%	41.3%
Spain	7.2%	31.2%	61.7%
UE-15	4.5%	29.2%	66.1%

the 'leisure civilisation' in the contemporary developed world led policy-makers to the prioritisation of tourism promotion.

To pass from the image of a devastated industrial port city to that of a tourist resort of national and European standing is an enormous challenge for an urban community, even allowing for the positive attitude of the responsible public bodies.[3] Regeneration has now been under way for a little over 15 years. In reconstructing the history of the city's growth and the evolution of its urban form, it appeared interesting to examine the bases on which this functional redeployment was founded. Was it a matter of wiping out an industrial and maritime past that was seen as negative for the competitive promotion of the city at national and European levels? Or were efforts made to promote interaction between industrial traditions, redefined to engage with the future, and the construction of an innovative city of culture, in which the heritage and cultural memory of the location was called upon to accompany and legitimate the strategies of innovation, with participative and educational goals aimed at gathering popular support and attracting visitors from other places?

Urban Morphology: Two Centuries of Industrial and Seaport Functions

For many years, Gijón remained a small seaport of regional significance on an ancient site which had been occupied by the Romans in the 1st century A.D. In competition with Avilés during the early modern period, the town had not grown beyond 6000 inhabitants at the end of the eighteenth century. Its growth was stimulated by the first industrial revolution and the presence of coal mines in its hinterland, although colonial links with Cuba also promoted industry through the establishment of a tobacco factory in 1822. Rivalry between agricultural and mining interests among the local nobility delayed the creation of the 'coal road' from Langreo, which was completed between 1838 and 1842, but it was the railway, inaugurated in 1856, that ensured the rapid development of the port of Gijón. The combination of Asturian coal and Basque iron ore provided a solid basis for the development of metallurgical industries.

Adapting the port to the needs of the regional economy proved a complex and contested process because of the opposition between supporters of the expansion of the old harbour and advocates of a new site at the western edge of the bay.[4] The royal decree of 30 October 1891 endorsed the latter scheme by approving the construction of the Musel, which at last allowed contracts to be issued in 1892 after 15 years of fierce debate.[5] The supporters of the expansion of the old harbour had for a long time included the Town Hall, the elite social club, the governing body of the port, the naval command, the merchants, the local newspaper *El Comercio*, several wealthy urban property owners and most local businesses. The 'Muselista' party brought together, above all, the industrial and financial *bourgeoisie*, particularly the steel manufacturers and the mine operators, but also the owners of estates backing on to the western side of the bay.

Throughout the nineteenth century, during which the town's population grew to around 47,500 people, the expansion of the urban area was checked by the ramparts erected during the first Carlist War of 1833–1839. The proximity of the port and the coming of the railway promoted development to the west of the town, following the bay and mingling industry and workers' housing, although from the 1860s some of this development was diverted to the other side of the central promontory, along San Lorenzo beach.

This plan of the town and port of Gijón (Figure 10.1), drawn up just before the great conflict began, identifies all the locations mentioned in this chapter. We can see how the urban area opens out to the south from the streets of the medieval and early modern 'old town' nestling at the foot of the peninsula of Cimadevilla but still confined within the walls. The new town follows the kind of grid plan that was common in cities across Europe from the eighteenth century onwards. The original harbour has a sheltered location to the west of the promontory. Industry is developing on the land behind Natahoyo beach, transforming this suburb into the first working-class quarter of Gijón.

This hierarchical arrangement, on to which was grafted the main axes of road and rail communication with the centre of Spain, continued to condition the town's expansion during the first half of the twentieth century as the population reached 101,000 inhabitants in 1940.[6]

Spain's loss of Cuba in 1898 made important investments available for redeployment in constructing the new port and associated activities.[7] The allocation of transatlantic trade with Latin America, inaugurated by the German steam packet *Santos* in 1910, helped to stimulate the port's economy. In 1924, Gijón was the fourth most important port for emigration

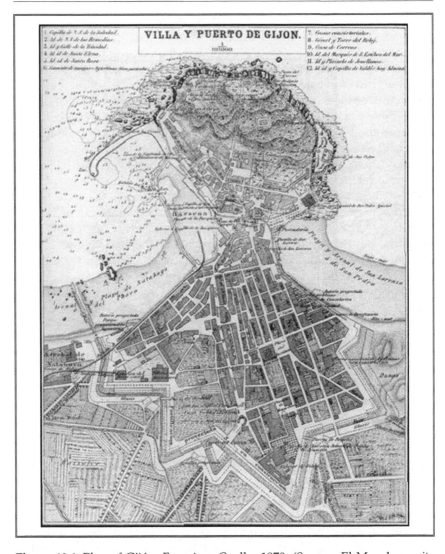

Figure 10.1 Plan of Gijón, Francisco Coello, 1870. (Source: El Musel, *op. cit.,* p. 60.)

to Latin America, behind Vigo, Corunna and Barcelona.[8] At the end of the Second World War, Gijón had reached the first rank of Spanish commercial ports, with 3 million tons of traffic. In 1945, the port authority began discussions on expanding the installations, producing a set of outline proposals which formed the basis for all later initiatives. In 1946, a new regular service to the Caribbean (Venezuela, Puerto Rico, Cuba) was added to the one already inaugurated in the 1930s to Havana and New York, strengthening Gijón's role as a transatlantic mail port.

Gijón's urban landscape experienced a complete upheaval during the 1960s and 1970s, which saw the most rapid demographic growth in the city's history, from around 125,000 inhabitants in 1960 to 256,000 in 1981.[9] This rapid doubling of population was accompanied by the enfeeblement of local government and the deterioration of the urban landscape. The continuation of older modes of operation had previously allowed the municipal government to cope reasonably well with slower growth following its urban plan of 1947, which had enabled it to embellish and sustain such assets as the marine boulevard San Lorenzo and the extensive park of Isabella la Católica, which had been reclaimed from marshland at the eastern edge of the city.

Figure 10.2 shows the three harbour basins which developed from the expansion of the old port. The oldest provided anchorage for only the smallest vessels, most of which were used for inshore fishing. Some medium-sized ships, like the one that is shown leaving port, still made use of the port's services, which remained railway-based. But the old coal traffic had been completely transferred to the Musel. On the second quay which separates the two most recent docks, there are buildings which seem to be stores and ship repair workshops. On the outside quay which closes off the old harbour, facilities for loading and unloading are still operational.

Figure 10.2 Aerial view of the docks of the old port, Gijón, towards 1960. (Source: El Musel, *op. cit.*, p. 119.)

The changes that are visible here arose from the opening of a new cycle of expansion based on the steel industry, which gave a formidable boost to all aspects of the economy, including the improvement of the port's infrastructure,[10] along with a fundamental upheaval in the structure of the port's traffic, because it was now necessary to import coal to meet the needs of the new steelworks. The extension of the Musel harbour works was a response to this powerful economic expansion. In 1985, the national plan for industrial coal imports divided the traffic equally between northern and southern Spain, concentrating northern imports on the Musel.

This second great phase of industrialisation caused explosive growth in the labour market and stimulated strong migration flows into Gijón, above all from rural Asturias and from the declining coal mining districts of the region. Local government proved unable to cope with the problems presented by population growth on such a scale. The combination of the prevailing tendency towards laissez-faire and the urgent need for responses to an explosive growth in social needs brought about the subversion of public controls by private initiatives. The sheer scale of the problems, which required rapid solutions, led to accelerated and chaotic development in which speculative interests soon took precedence over public regulation.[11] This was the period, above all, in which planning norms were set aside. This gave rise to the severe overcrowding of the old city centre, which saw the disappearance of a significant number of historic buildings. There was also a rapid extension of new development along San Lorenzo beach, together with the degradation and accelerated proletarianisation of the old fishing quarter at Cimadevilla, and the very rapid expansion of the western working-class districts and of the nearer outskirts of the city, as the provision of basic public services failed to keep pace with speculative growth. Worst of all was the development of shanty towns on the urban fringe or on abandoned industrial sites.

Gijón was then hit very hard by the industrial crisis which affected so many port cities during the 1980s and 1990s, with particularly severe consequences in this case because of the cumulative importance of employment in coal, steel, shipbuilding and textiles. This sharp contraction of industrial activity occurred under adverse social conditions, with radical workers' struggles. But this did make possible the reformulation of a positive urban planning policy, emerging from a redefinition of the vital functions of the city and of its regeneration through the reallocation of industrial resources. The first stage was associated with the adoption of a new urban plan in 1986, which based its legitimacy on an energetic reaction from public authorities against the outcomes of uncontrolled expansion which had led to the heightened degradation of the urban environment and a severing of social bonds within the community.

A new stage of development was identified with a new dynamic of urban change, as much economic as social and cultural,[12] through the mobilisation of interested parties behind a strategic development plan whose first incarnation ran from 1991 to 1999.

This first experience of large-scale urban regeneration provided an apprenticeship and a point of reference for the construction of the second (current) plan, covering the decade 2002–2012. This plan, the outcome of an important democratic debate organised by the city government, is articulated through 10 major themes which are intended to interact and sustain a new urban dynamic. Certain themes are seen as basic, while others, such as themes 2, 4 and 6, are intended to cut across and make links.

Official presentation of the Gijón strategic plan, 2002–2012[13]:

(1) Active and entrepreneurial city
(2) City of educational excellence
(3) Centre of cultural tourism in the Atlantic arc
(4) City of innovation
(5) Logistics centre of the Cantabrian coast
(6) City of sustainable development
(7) Secure and accessible city
(8) City of equality and social cohesion
(9) City of participation
(10) Municipal government

The urban regeneration drive thus rests on three fundamental supports: a good employment level for the creation of numerous lasting enterprises; the endorsement of a cultural and touristic orientation as a powerful motor of growth and endorsement of the transformation of the city's image; and an aspiration to play a major role in communications across the north of Spain. It is through the increasing interconnection of these three bases that Gijón proposes to utilise its character as a port city and its identity as a European Atlantic coastal city to create a positive image for itself in competitive international markets.

The economic regeneration has not affected the trajectory of the port of Gijón, which has experienced a very slow but steady rise in its total traffic since 2000. It is identified above all as the most important Spanish port for the importation of solid products, particularly of high-quality industrial coal and of iron ore.[14] The city's advertised ambitions towards becoming the main seaport of north-west Spain, serving principally the autonomous regions of Castilla-León and Madrid, have even justified the preparation between 2001 and 2004 of a major investment plan to extend the site of the Musel. The project that was adopted is based on a financial provision of 579.2 million euros, of which 247.5 million were made available by the European Union. Construction began in February 2005.

Gijón, Centre for Tourism and Culture of the Atlantic Arc

The deep affinity between the city and the sea is reaffirmed by the Strategic Development Plan's attachment to tourism and cultural activities as the second decisive motor of urban renaissance. A quarter of a century ago, it needed all the innate optimism of a tourism office to present Gijón as a major tourist staging post that might justify a detour on a longer trip or an extended summer stay. The port and industrial activities had completely swallowed up the extensive bay to the west of the central peninsula, largely cutting off the city from the sea and making the beaches vanish under industrial plants. On the other side, all along San Lorenzo beach, insufficiently controlled and ill-considered urbanisation had allowed the construction of an enormous concrete barrier along the boulevard, where a heterogeneous built environment devoid of aesthetic consideration could only prompt feelings of revulsion at the degradation and premature ageing of façades exposed to salt-laden winds. We should add that the great height of the buildings prevented the sun's rays from warming the promenaders or gilding the bathers on a north-facing beach. In laying out its eastern seafront, Gijón made the mistake, now difficult to put right, of allowing the construction of a wall of particularly unpleasant aspect, an automatic source of negative opinions about the city's image, its local government and even its inhabitants.

In the most central district, the old fishermen's quarter of Cimadevilla, built at the foot of the fortified peninsula, had fallen into terrible disrepair as its unhealthy state led to social marginalisation. The heart of the city, concentrating essential retailing and other central services, had become almost unreadable despite its attractive examples of modernist architecture from the early twentieth century, since pollution had covered the façades with an obscuring veil of blackness. All these blemishes created an image of a port city in a state of serious degradation, hardly suitable for sustaining a high-class tourist presence. This is why the inclusion of tourism and culture as resources in the third level of the plan, as the second pillar of recovery, was a particularly strong demonstration of a will to succeed. But a visit to the Gijón of today will show that this was neither utopian nor unfounded, for the record of successful developments is already quite impressive.

As an ancient city, even though it remained small in size until the beginning of the nineteenth century, Gijón has been able to pull together a certain number of heritage assets, beginning with the remains of its Roman baths, to support its pretensions to cultural tourism through its historical monuments. Everything that was located in the very old urban core of Cimadevilla has thus been restored and recognised, such as the town house of the Jovellanos family, commemorating the minister and emblem of the Spanish Enlightenment who was the pride of the city in the late eighteenth century. The big houses of the nobility are today art galleries or exhibition centres. There is still ample scope for making full use of the most remarkable buildings of the city centre, constructed by businesses, banks and the commercial *bourgeoisie* at the turn of the nineteenth and twentieth centuries, or at least those that survived the destructive urban renewal of the 1960s and 1970s. The rehabilitation of the oldest part of the San Lorenzo seafront boulevard gives an indication of the way forward and the satisfying results that are obtainable.

All the same, the most powerful opportunities in Gijón are to be found not so much in ancient monuments as in the city's renewed embrace of the sea, in a new alliance that places all observers, whether inhabitants, holidaymakers or birds of passage, in a position to understand the city from a coastal vantage point, a privileged position shared by very few cities across the world. It is the accumulation of remarkable sites, of attractive leisure provisions and of contemporary artworks along an immense maritime promenade of several kilometres, that allows the multiplication of angles of vision towards the city, benefiting from the effects of perspective and the changes of relief with the play of capes, bays and the central peninsula, that enables an overall apprehension of the city's urban form.

At the nearest point to the heart of the city, the old coal harbour has been converted into a marina for pleasure craft. The old customs house, which was more recently a fishermen's storehouse until 1987, has been restored in a way that respects its traditional architectural style. Its central pavilion now houses the offices of the marina, with a seafood restaurant in one of the wings and an exhibition hall in the other, managed by the port authority of the Musel, which digs into its rich documentary and photographic archives to present thematic exhibitions. The site thus brings together authenticities of place and historical record to communicate the maritime heritage of the port of Gijón, splendidly located at a necessary place of passage for all visitors to and strollers along the quays of the old port, at the point of intersection between the two great stretches of the very extensive marine promenade and several pedestrian routes into the heart of the city.

The links between city and port are thus revitalised in three dimensions: through their inscription in a real urban centre apprehended physically by the pedestrian, through the expression as heritage of the original founding activities of the first phase of urban growth and through the city's renewed life as another way of using marine resources through leisure provision. The tourism office reigns in its central position beside the leisure services but also next to the maintenance workshops for pleasure craft. On the western fringe of the old port, a major thalassotherapy centre has been constructed, again returning to the use of the sea in ways that are adapted to new forms of consumption bound up in the development of the civilisation of leisure and well-being in countries with high living standards.

Figure 10.3 must date from the mid-1990s because, although we can clearly distinguish the two reconstituted beaches, it shows neither the great aquarium nor the thalassotherapy centre, but the Chillida sculpture (1990) is already in place. Between the two beaches, the survival of a shipbuilding industry, still active in 2009, preserves direct testimony of the industrial history of a site whose reorientation towards residential and tertiary sector functions seems to have just begun. At the back,

Figure 10.3 Aerial view of the urban renovation of Gijón's west bay, *c.* 1995. (Source: El Musel, *op. cit.*, p. 125.)

behind San Lorenzo beach, we can observe the barrier of very high buildings that presents such a problem for the current image of the city and its tourism promotion. The very tall buildings and excessive density at the heart of the city are also readily apparent. The old port has been converted into a pleasure harbour.

Moving westwards, on the land abandoned by the now shrunken shipbuilding industry but clearly visible at the back of the photograph, public access to the shore has been restored in the form of two old beaches – completely restored after being destroyed by twentieth-century industrialisation – and an extensive esplanade where people go to and fro using all kinds of non-motorised transport. Here, renewal has aimed at a mix of functions over and above the classic regeneration model through cultural leisure by including important public services – such as the law courts, which have been installed in an architecturally appropriate building – as well as residential provision. It is worth drawing attention to the architectural symbolism imposed on the developers of this seafront housing because it entails an explicit evocation of the shapes of vessels, with the intention of sustaining place-memory through architectural forms that recall a ship under construction on the exact site of the old shipyards. The renovation scheme as a whole is closed off by an enormous aquarium, an establishment dedicated to the display of the world of the sea, in both material and virtual form, through a strong multimedia presence and an emphasis on advanced technology. From the vantage point of its panoramic terrace, visitors have the opportunity to read the shape of the city and to distinguish the strong elements of regeneration which they may have the opportunity to explore.

On the east side, the ambition of promoting Gijón as a high-class tourist destination rests on four major programmes in which the concept of an Atlantic city remains predominant without being exclusive. On the way out of the city, the first task, at once enormous in scale and requiring delicacy of treatment, is of course that of disguising the 'wall of shame' formed by the imposing barrier of tall and charmless buildings erected less than half a century ago along San Lorenzo beach. From an aesthetic point of view, and to restore sunlight to the beach, the most obvious solution would be wholesale demolition, but such a radical approach has not been deemed realistic. The solution adopted has entailed external beautification through a decorative treatment of the façades, resting on the harmonisation of spaces through work on the balconies and the installation of a translucent film designed to disguise the mediocrity of most of the buildings. It is still too soon to judge the extent to which the project, the first sections of which have just been opened, has achieved the goals set out by the city architects and planners.

The next stage, already validated by visible public demand, is the extension of the San Lorenzo promenade towards the eastern fringe of the city, moving progressively from the urban to the rural and keeping close to the sea for the five kilometres of managed coastal path. This project has received immediate public endorsement because it recovers the link between sun and sea, whose proximity is experienced both in vision and sound, for this is a splendid coastline of sharply defined cliffs which multiply the viewpoints over the city while confronting the endless waves of the ocean.

In the same direction, but a little way back from the coast, on the eastern edge of the urban area, an immense botanical garden has been established, giving priority to Atlantic plant species. Pedestrian access allows visitors to grasp the extreme diversity of vegetation to be found in the Atlantic world, and an interpretation centre with modern visualisation techniques assists the historical understanding of the formation of the landscapes of Asturias and the Cantabrian coast as a whole, from a geological as well as a botanical perspective.

Forming a powerful cluster of visitor attractions, the same zone is the site of the Technical University, erected by the Franco regime in the 1950s. This is an imposing, monumental architectural composition, originally dedicated to technological education, and the city is now reconsidering its role. Starting from an awareness of the recognised monumental attractions of one of the most important Spanish architectural creations of the mid-twentieth century, the city is planning its transformation into the crowning glory of its national and international tourism reputation. Instead of recruiting an architect of global renown to create a flagship project as the symbol of an urban renaissance, as in the case of many port cities throughout the world, such as Bilbao and before long in the competing Asturian city of Avilés, the city government has concluded that the functional redefinition of a great heritage building, together with its internal rehabilitation, could compete very well with any contemporary creation. We should underline the originality of this approach in contrast with the marked tendency in port city regeneration towards 'waterfront' programmes which tend to marginalise heritage buildings for fear of nullifying the desired image of reasserted modernity, which has been an absolute priority in the promotion of urban branding in a context of national or international competition.

The strategic plan for 2002–2012 thus presents the Technical University as 'the great cultural, educational and touristic project which will entail the creation of a shared multicultural space'. The revitalisation of a group of monumental heritage buildings does not neglect contemporary creativity, which can be found at the very heart of the construction of a new image of an innovative and entrepreneurial city according to the plan; nor does it ignore the power of historical memory through the mobilisation of the city's industrial tradition as a support for the future.

The position adopted by D. Juan Cueto Alas, vice-chair of the planning, infrastructure and environment axis of the plan, at the final presentations of the strategic plan in November 2002 bears eloquent witness to this:

> The present project for the reclamation of the Technical University symbolises the spirit of combining the different themes within which the future of the city has been inscribed and which cut across this Strategic Plan. In the first place, architectural regeneration, and a utilisation plan adapted to current needs (from traditional vocational learning to the development of new university faculties) and then, at the same time, the creation of the Centre for Contemporary Art and Industrial Creativity, based on a combination of information technology and the use of the products of regional industries (steel, aluminium, glass, etc.) in the diverse artistic creations of the twenty-first century. An authentic city of culture, rooted in the industrial flesh and bone of our region, which would attract an important flow of cultural tourism, which would connect us with avant-garde cultural centres in Europe, and would locate us within the important cultural axis which is emerging on the Cantabrian coast (from the Bilbao Guggenheim to the future Cultural City of Santiago de Compostela).

The combination of powerful elements destined to construct a tourist attraction on a national and European scale must find its cohesion through the affirmation of the city's Atlantic identity, while engaged in the change from an industrial to a service economy in which tourism must become one of the leading drivers of development.[15] Specifically, the geographical dispersion of sites, monuments and attractions along an extensive seafront is counterbalanced by its insertion into an integrated network of traffic flows, providing the opportunity for a global extension of this urban phoenix while its central arc can easily be traversed on foot. Two contrasting urban promenades thus converge on the axial promontory of Cimadevilla, the historic site of the city's foundation, reorganised in its upper area in consonance with its historic coastal fortifications in an impressive balcony of greenery which overlooks the ocean to the north and the city as a whole to the south. The summit of the promontory is crowned by the imposing contemporary concrete sculpture by the famous Basque artist Eduardo Chillida, a grandiose work entitled 'Eulogy to the horizon'.

Over and above the power of its symbolism of opening out towards the extensive; the external, the exploratory, and the rejection of frontiers implied in the alchemy of form, place and monumentality, the main interest in this sculpture arises from the variety of scales of vision that are brought back to the individual dimension. At the foot of this work, the observer is impressed by the imposing grandeur of what is still a city gateway between earth, sky and sea, or by how the ship's figurehead of the new Gijón is ready to set off in search of new horizons; however, the further the observer moves from the construction, the more firmly based in the general setting of the city is its imposing mass, while remaining immediately identifiable. Additional contemporary sculptures, in very varied styles,[16] jostle with each other along the marine promenade and link it with the routes into the heart of the city, which contains its own public ornaments.

Figure 10.4 emphasises, from a different angle, the main under-pinnings of the promotion of Gijón as a centre of tourism and culture for the Atlantic arc. Centred on the promontory of Cimadevilla, it gives pride of place to the promenade that leads to the monumental Chillida sculpture, the focal point of the coastal route through the city. In the west bay (to the right of the photograph), the pleasure harbour stands out, as does the residential district which has partially replaced the shipyards. On the eastern side, we can see the start of the wall of high buildings along the San Lorenzo boulevard, casting its shadow over the beach of the same name.

Gijón became an important city in the nineteenth and twentieth centuries, basing its growth on its roles as a seaport and industrial city in

Figure 10.4 General view of the city, seen across the promontory. (Source: El Musel, *op. cit.*, p. 126.)

Asturias and Spain. The history of the relationship between the city and its port follows a standard evolutionary trajectory, with a divorce in the twentieth century occasioned by the creation of the new harbour of the Musel at the west end of the bay; however, it is also distinctive through the persistence of port activity on the original site, very close to the heart of the city. This industrialisation through two great historical periods has profoundly modified the urban landscape, at times for the worse, as in the urban overspill of the years between 1960 and 1985. The deep crisis in the steel, metal, shipbuilding and textile industries left its legacy in the area of the old port but also provoked a lively and ambitious reaction from local politicians who were determined to revive their city. Without neglecting industrial revival, a reorientation towards the service economy justified the promotion of the city as a cultural centre and key point of reference for tourism in the Atlantic arc. After a quarter of a century, the transformation of urban space testifies to this determination, particularly through the changes to the landscape of the seafront. This orientation to tourism was not opposed to the seaport dimension of the Atlantic city because this was, in turn, reinforced in triplicate by the great extension programme for the port of Musel, the success of the marina for pleasure craft and the development of maritime and industrial heritage through the celebration of collective memory and the promotion of cultural tourism in association with contemporary art.

Notes and References

1. Plan Estratégico de Gijón, Ayuntamiento de Gijon, 2002, p. 49.
2. M. Llorden Miñambres, *Desarollo económico y urbano de Gijón, siglos XIX y XX* (Oviedo, Universidad de Oviedo, 1994).
3. L. Suárez Fernández, *Reflexiones sobre la Historia de Gijón* (Salinas, Ayalga Ediciones, 1995).
4. After a brief republican interlude in 1873, punctuated by the outbreak of the third Carlist war and a succession of coups d'état in 1874, Canovas was able to restore the monarchy under Alfonso XII in 1875 and secure a new constitution in 1876, which was much more liberal in its inspiration, basing the stability of the new regime on the alternation in power of a conservative and a liberal party, each loyal to the Crown, and working through a system of local political manipulation.
5. F. Erice, *Propietarios, comerciantes e indusriales. Burguesía y desarollo capitalista en la Asturias del siglo XIX (1830–1885)* (Oviedo, Universidad de Oviedo, Servicio de Publicaciones, 1995).
6. J.L. Carmona García, 'La ciudad y el Puerto de Gijón (1900–1950)', en El Musel, op. cit., p. 83–102.
7. M. Álvarez González, *Historia de la obra pública local en la Villa y Puerto de Gijón. La obra pública portuaria (1750–1950)*, unpublished monograph, Ayuntamiento de Gijón, Autoridad Portuaria de Gijón, 2003.
8. The vessel of the Spanish transatlantic steamship company, the *Cristobal Colón*, provided a regular service to Havana, Veracruz and Tampico, and

embarked 900 emigrants on 20 December 1924, a record for the port of Musel.

9. Plan Estratégico, Ayuntamiento de Gijón, 2002, p. 68. In reality the growth of the city itself was much faster, passing from 90,000 to 233,000 inhabitants, while the population of the rural area of the municipality fell from 32,700 to 23,400 inhabitants.

10. R. Alvargonzález Rodríguez, *Industria y espacio portuario en Gijón* (Gijón : Junta del Puerto, 1985), 2 vols.

11. F.J. Granda, Gijón 1950–1986 : Un paisaje urbano en transformación », in El Musel, op. cit., p. 105– 127.

12. *De tu Historia. Gijón, 1937–1977. Sesenta años de ciudad.* Ayuntamiento de Gijón, Catálogo de exposición, 1997.

13. Plan Estratégico de Gijón, op. cit., p. 13–22.

14. Memoria Anual, 2005. Puerto de Gijón, pp. 18–22.

15. Plan Estratégico, Ayuntamiento de Gijón, 2002, p. 203. Presentation by D. Alonso Puerta, vice-chair of development axis 4 : Participation and civic image. 'The European context gives clear importance to the role of cities, very often organised in networks for common interest or solidarity... European cities are veritable centres of innovation, communication and creativity... I affirm my agreement with the commitment and willingness to promote the position of Gijon at the core of its geographical and economic setting in the Atlantic arc and the Cantabrian coast...As to the external visibility of the city, it is vital to belong to a network of cities, and in our case the situation is excellent with the participation of Eurocités, l'Arc Atlantique, Cideu, Leda, and the Kaleidos network...Gijón is a maritime city with a changing level of animation according to the seasons, but it is relevant to underline, as this document shows, that it is a lively city throughout the year.'

16. For example, a group of large steel plates features at the urban gateway to the city, as an evocation of the memory of the industrial city. Another, figurative female statue calls up the historic importance of migration through the port.

Chapter 11

From a Baltic Village to a Leading Soviet Health Resort: Reminiscences of the Social History of Jurmala, Latvia

SIMO LAAKKONEN and KARINA VASILEVSKA

Today, the resort of Jurmala is a small city of 55,000 people in the Republic of Latvia, which has 2.3 million inhabitants in all. The port of Riga, capital of Latvia, is located only about 20 kilometres from Jurmala. The history of health resorts in the Baltic Sea Region has been examined on a general level, but there are few in-depth studies.[1] In this chapter, we will examine the relationship between Riga and Jurmala. We aim to reconstruct an overall picture of Jurmala's development from the eighteenth to the twenty-first century, but our focus will be on the period from 1945 to 1991, when Jurmala developed into one of the leading resorts in the Soviet Union.[2]

Our overall description of the development of Riga and Jurmala before the Soviet period is based on literature and some archival sources. Our main source for the Soviet and post-Soviet period is ordinary people's reminiscences of the social history of Jurmala and Riga. Oral history is of great importance in studying areas of the former Soviet Bloc because hardly any other sources are available that would reflect the viewpoints of the majority of the people. In this study, semi-structured interviews were conducted with 14 people living in Jurmala.

Jurmala during the Russian Empire, 1700–1914

From the beginning, trade and militarisation shaped Riga's form and history, thus affecting future possibilities for recreation in and near the city. Originally, the current location of the capital city of Riga was a settlement of a Finno–Ugrian tribe, the Livonians. Germanic crusaders founded the City of Riga as a military centre in 1201 to subdue the local Livonian and Baltic tribes. Later, Polish and Swedish invaders were also interested in Riga as a military site.[3] Consequently, people lived very densely within the city walls of Riga. The risk of epidemics was high, and in 1848 cholera claimed 7000 lives, killing 1 in 10 of the inhabitants.[4]

Riga is not situated on the coastline but about 15 kilometres from the Bay of Riga, an arm of the Baltic Sea. It became an important port city when it entered the Hanseatic League of German trading towns at the end of the thirteenth century. Because of the increase in the size of the ships and the shallowness of the River Daugava, the port facilities were gradually expanded towards the mouth of the river. In the nineteenth century Riga was, after Archangel, the second most important seaport of the Russian Empire, and consequently an increasing area of both banks of the River Daugava was occupied by harbour premises.[5] Another obstacle for resort activities near Riga was the 1000-kilometre long River Daugava itself. The risk presented by the horizontal movement of the powerful river, pack ice, floods, and also related hydraulic engineering constructions hindered recreational use of the river near the city.

Riga and Jurmala and the territory of contemporary Latvia became part of imperial Russia in 1721 at the end of the Great Northern War. The health and leisure potential of Jurmala, which led to its later development into a resort, was already noticed at the end of the eighteenth century when the first visitors started arriving during the summer. This follows the general trend all over Europe, inspired by Romanticism and the preference for nature-based recreation in areas that were previously considered hostile and inhospitable such as coastlines and mountains.[6]

Before the arrival of the first summer visitors, the territory of today's Jurmala was deemed infertile for cultivating and was thinly populated by fishermen. The area had no port or manufacturing facilities of importance. The specific natural resources – such as seaside, pine forest, hydrogen sulphide and other mineral water springs and nearby Kemeri mud – played a crucial role in the historic development of Jurmala into a health resort (see Figure 11.1). The first state-run medical sanatorium with 20 tubs was opened in Kemeri in 1838. During the cholera epidemics that hit Riga in the 1830s and 1840s, Jurmala was also sought out as a safe haven.[7] The contrast between the contaminated port city and healthy resort was already clear by the early nineteenth century.

The picturesque wooden summer cottages that serve as the visual symbol of Jurmala today were mostly built at the end of the nineteenth century and beginning of the twentieth century.[8] Plots were rented out by mostly Baltic German landowners to the Russian, German and wealthy Latvian summer guests. The majority of permanent residents were Latvians who earned a living in fishing and agriculture, but they were outnumbered during the holiday season. The railway connected Jurmala to Riga in 1877 and soon also to the main cities of Imperial Russia, from where an increasing number of holidaymakers arrived.[9] In the period between the end of the nineteenth century and the First World War, Jurmala grew into a health resort with a flourishing cultural scene known throughout the Russian Empire. The number of summer visitors

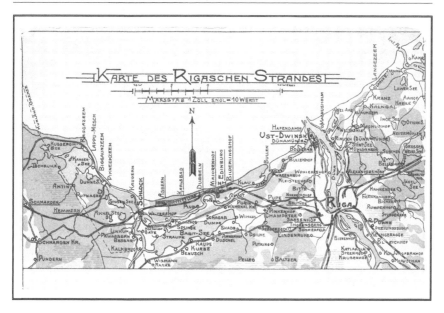

Figure 11.1 German map presenting the different beaches of Jurmala resort, early 1910s.

increased from 18,000 in 1864 to 80,000 in 1914.[10] Famous writers such as Latvian national poets Rainis and Aspazija and the Russian author Maxim Gorky, together with other artists, wealthy bureaucrats and Tsarist officials, spent their summers in Jurmala (see Figure 11.2).[11] As in Swansea, examined in another case study in this book, Riga and Jurmala achieved a certain symbiosis through the zoning of commerce and leisure.

Latvia's First Independence, 1918–1940

Because of the liberation of serfs and rapid industrialisation, the population of Riga increased quickly from almost 200,000 in 1881 to over 500,000 prior to the First World War.[12] Industrialisation transformed Riga from a trading town to a major manufacturing centre. Riga became a cosmopolitan city and, located on the extreme western edge of the Russian Empire, the window to the West across the Baltic Sea, as well as the centre of growing Latvian national awareness and political activity.[13]

Latvia gained its independence from the Russian Empire in 1918. At that time, Jurmala consisted of several bathing unions (Badvereine) with independent local administrations. A union of resort territories obtained city status for Jurmala in 1920 when it was named Rigas Jurmala. Kemeri and Sloka, which are now part of Jurmala city, were both independent towns during the inter-war period. Kemeri continued to develop its

Figure 11.2 Dr Maximowitsch's water treatment establishment/sanatorium was situated among the pines of Edinburg beach when Jurmala was still under the rule of Imperial Russia.

facilities with new restaurants, parks and health services, capping it with the building of a lavish hotel (1936) that is one of the best architectural examples of the inter-war period.[14]

Jurmala began to flourish again by the late 1920s and early 1930s as more and more people wanted to get away from the discomfort of the growing industrial centre of Riga. Jurmala gained a much more Latvian character both socially and visually over this period. The visitors from the newly formed Soviet Union ceased coming – thus initiating the only time in Jurmala's resort history when it was not visited by holidaymakers from Russia. As before the First World War, Jurmala was a resort for the wealthy upper and upper middle class inhabitants of Latvia (see Figure 11.3), but it was not a luxurious resort with overly expensive hotels and entertainment; rather, it was a place for many writers and artists to spend their summers. As before, it was also a scene for cultural activity such as concerts and poetry readings, but they had turned much more Latvian in flavour. The growing Latvian working class, however, did not spend their holidays in Jurmala.

The lives of the local inhabitants of Jurmala, many of whom still worked in fishery and agriculture, but some of whom had completely turned to the tourist industry, went side by side with those of the summer visitors, rarely intertwining, very much as had been the case before the

Figure 11.3 Bathing establishment in Bulduri, Jurmala, in the 1920s, indicating that the resort was mostly used by middle and upper class people.

First World War. During the inter-war period, it was quite common for owners to be living in one house and to have several summer houses built on their property to host the summer visitors, who rented a house, not separate rooms, as was to be common during Soviet times.[15]

Constructing a Soviet Health Resort

The independent Republic of Latvia was formally annexed into the USSR in August 1940. In the following year, the Latvian Soviet Socialist Republic (SSR) was conquered by Nazi Germany before being retaken by the Soviets in 1944–1945. Jurmala was only partially damaged during the Second World War, and it welcomed the first holidaymakers as early as the summer of 1945. Approximately 30,000 visitors came in 1946, and by 1959, this number had reached 107,000.[16]

Jurmala remained popular for several reasons. Riga expanded and industrialised rapidly during the socialist era. The pre-war gross industrial output of Riga in 1940 was exceeded by a multiplier of 13 by 1960 and 48 by 1980.[17] A high number of guest workers were transferred or moved from other republics of the USSR to Riga. With this, the number of inhabitants of the capital city increased from the pre-war figure of nearly 400,000 to almost one million in 1989. The military status of Riga was revived because of the Soviet occupation, and it became the headquarters of the Baltic military district. Many of the blank spots on official plans were industrial districts, transportation terminals, and sites for military training. In turn, it was widely understood that recreational activities were not expected to take place around Riga.

Jurmala's network of sanatoriums and other medical services was placed under the jurisdiction of the Ministry of Health in Latvia in 1956, which can be seen as the first step towards the planned development of Jurmala into a medical resort. City status was bestowed by uniting the Riga Jurmala district with the towns of Sloka and Kemeri in 1959. The main aim of the newly formed Jurmala City was to turn it into a major health resort to serve the whole USSR. Arnolds Deglavs, who became the head of the new City Executive Committee in 1961, was the main visionary and promoter of the post-war development of Jurmala.[18] He had allegedly visited French and German resorts such as Baden Baden and was known to have expressed his wish that Jurmala be like them.[19] As a result, the city of Jurmala, 'Riga seacoast' (*rizhskoje vzrmorje*), was officially transformed into a healing space (*zdarvnica*).[20] This scheme was finally approved in 1971 by the Council of Ministers of the Republic, and the new city plan was based on the three priorities of promoting health, resting and sport.[21]

One consequence of the plan was that the national and local train connections were greatly improved. The number of tracks was increased, and electrification was introduced. Finally, there were 120 train connections per day to Jurmala during the summer.[22] The first standardised blocks of flats were built in the late 1960s to house the ever-increasing number of resort workers.[23] Restaurants, cafeterias, concert halls, resting houses and medical sanatoriums were built in Soviet style to accommodate the ever-greater influx of visitors from the whole of the USSR (see Figure 11.4). On the other hand, unlike many traditional seaside resorts in Western Europe (see the chapter by Brodie in this book), socialist Jurmala protected its seaside from harbour and industrial activities. Also, a list of architectural monuments to be protected was approved in 1980.[24] The urban centres were also protected in the early 1980s by the introduction of a toll for incoming cars that remains in effect to this day.[25]

Kurtology (*kurortologija* in both Latvian and Russian), the scientific study of health resorts and healing processes, was taught at the University of Latvia,[26] and a scientific research laboratory for kurtology was established in 1962.[27] Jurmala became well known for treating patients with cardiovascular diseases. Rheumatism, movement and balance impairment, gastro-intestinal conditions and infertility were also treated.[28] In 1971, Jurmala was awarded the all-union status as a health resort city (*gorod-kurort* in Russian),[29] and it gained unprecedented recognition all over the USSR.

Those who came to Jurmala were divided into two groups, the so-called organised and unorganised visitors. The former received a *putevka*,

Figure 11.4 Mass tourism started in the 1960s and Jurmala became a major resort in the USSR.

or voucher, from their trade union, which included meals and medical treatment, if this was prescribed, and travel costs were mostly covered.[30] Unorganised vacationers rented rooms directly from the residents of Jurmala. The regular price was a rouble for a bed per night.[31] The rental process was supposed to be highly supervised, but many visitors pretended that they were relatives of the host. Moreover, the mass of the holidaymakers was too much even for the Soviet security system to control. The vacationers came from all parts of the former Soviet Union but most often from Moscow and Leningrad.

Jurmala became a stylish resort popular with the Soviet elite, artists, intellectuals and high-ranking officials from all over the USSR. The local leaders of the Communist Party were keen on hosting their influential visitors. Two elite sanatoriums that were meant for the top party officials were built.[32] The USSR Council of Ministers leisure home – or Kosygin's house, as it was called by the locals – was completed in 1972.[33] The party secretaries Nikita Khrushchev and Leonid Brezhnev are known to have frequented the resort incognito, as did Boris Yeltsin later on.[34]

About 300,000 organised holidaymakers and 40,000 to 50,000 unorganised holidaymakers came to Jurmala every year in the early 1980s. On a sunny summer Sunday, as many as half a million people could gather in Jurmala, turning it into a socialist Baltic Riviera.[35] Many interviewees remember that, during sunny weekends, it was impossible to board the overcrowded trains, and there was a never-ending stream of visitors.[36] Jurmala became the third biggest health resort of the USSR; only Yalta and Sochi by the Black Sea were larger.[37] Mass tourism to Jurmala was based on the fact that, in the socialist system, besides the elite and the so-called Soviet middle-class, the better-off working class for the first time enjoyed the possibility of engaging in leisure and tourism. This was a major change in the life of the people in Latvia and in the USSR in general. From this social perspective, the coast of Jurmala became more democratic.

Latvians and Jurmala in the Soviet Era

The Soviet regime brought about significant changes in Jurmala. The number of inhabitants grew steadily between 1950 and 1989, from 35,000 to 60,000. The ethnic composition of the population changed radically because, before the Second World War, the majority of local residents of Jurmala (86.6%) were Latvians, but in 1989, more than two-fifths (42.1%) were Russians.[38]

Local residents were allowed to own the house where they lived, but non-residents' summer houses were nationalised and turned into 'communal houses' where each family was allocated one or two rooms. These often did not have running water or a toilet inside.[39] A municipal water supply and sewage system began to be built only in the 1960s.[40]

Local people tried to benefit from the growing flood of holidaymakers by creating new houses or rooms. One of the interviewees remembers that building a house in Jurmala in the 1960s was a constant struggle, because building materials had to be bought on the black market and were products of the fairly common *gosudarstvenaja krazha* (stealing from the state). Almost every local inhabitant rented out rooms. As one of the narrators recalls: 'It was crazy with the renting out back then. There was even an expression: "Shooed a goat out and let the renters in." We even built our house, thanks to the money we took in from renting out those parts that were finished.'[41]

Renting rooms, fishing and farming remained important sources of income for local residents. A chief doctor of a sanatorium recalls that Latvians were working 'like horses' so that everything would be well in the resort.[42] As a result of the ever-growing resort industry, the lack of staff is a recurring theme in the local press in the 1980s.[43] Sea and mass tourism benefited those Latvians living in Jurmala but not those outside the city. If they needed to go to Jurmala for medical reasons, they could do so only outside the summer season. In Jaunķemeri sanatorium, for example, only 3%–5% of the patients were from Latvian SSR.[44] However, most Soviet vacationers did not attend any of the medical treatments (except for the massages) and spent their time on the beach.[45]

Cultural Aspects of Soviet Tourism

As Anne Gorsuch notes in her account of tourism in the latter years of Stalin, the investment in recreational facilities owed much to their value as an ideological tool. The USSR was already entering the Cold War phase in its relations with the outside world and was turning inwards. Promoting domestic tourism was considered a good way of acquiring the much-needed Sovietness.[46] The inflow of the newly acquired fellow 'country-men' to Latvian SSR could be seen as an attempt at Sovietisation, but their encounters with the new republic can also be analysed as a manifestation of their perception of Pribaltika – that is, the three Baltic republics of Estonia, Latvia and Lithuania – as a 'Western' place.

Riga and Jurmala presented a Western lifestyle with strikingly different traditions, architecture, language, alphabets and ambience (see Figure 11.5). 'Latvia was considered the frontier-post of the Soviet Union,

Figure 11.5 Despite Soviet modernisation historic buildings such as this seaside hotel at Majori beach in 1980 remained an important part of the lure of the Latvian seaside.

the furthest point towards the West. It was considered the example of "culture".[47] While the capital of Latvia, Riga, was an ever-growing, industrialised local metropolis, it was Jurmala that came to symbolise this Western 'otherness' in the popular understanding of the Soviet citizens.

Consumer goods were also more accessible in Latvia than in many other Soviet republics. Such artefacts as amber or even plastic jewellery were made and sold by the private handicraft sector.[48] Galina Pagiraine remembers that in Jurmala it was possible to taste forbidden things, such as smoked fish, which was served in her favourite café and had to be eaten, literally, 'under the table'.[49] It was also considered that local café and restaurant culture was more relaxed, and the service standards were higher than in other republics.[50]

In Soviet ideology, 'the productive value of touring and travel was for intellectual and physical self-improvement'.[51] Therefore, sport and culture became an important part of Jurmala's self-image. Youth gymnastics was promoted.[52] Local tennis courts hosted Davis Cup matches from the 1970s. Sport boats were built in the area. Jurmala hosted a USSR-wide competition for new popular music singers.[53] Other concerts and excursions to museums and to the countryside were arranged. Also, unofficial entertainment was available, with a cabaret restaurant *Kaburgs* operating quite successfully. However, local residents attended tourist events only rarely.[54]

During the 1960s, the streets of Jurmala were widened at the expense of the pavements; fences that traditionally had kept the houses separate from the streets were taken down, and new sidewalks were put in the yards of the private homes. For Galina Pagiraine, who regularly went to Jurmala from Riga with her friends in the 1960s and 1970s, 'a promenade along the main street' gave a sense of freedom.[55] Inese Ābola remembers that after a long and hot journey in an over-crowded train, the coastal fresh, pine-scented air and the sea seemed very much like a paradise and was in stark contrast with her everyday life in Riga.[56] While political sentiments were kept hidden, it was clear that the seaside had become a symbol of freedom.

The New Era since 1991

The independence of Latvia was restored on 21 August 1991. For Jurmala, this change was particularly harsh.[57] A dramatic halt to tourism took place in 1993 when all foreign visitors, both from the east and the west, stopped coming. As a result, almost all of the 105 sanatoriums operational in Jurmala in 1980 were either closed or run down and sold off. Four sanatoriums remained in operation: one state-owned, one Belorussian, one Russian and one privately owned, Jaunkemeri. Galina Pagiraine recalls that Jurmala looked at that time like a totally abandoned place.[58]

Jurmala's tourism survived only thanks to the inhabitants of Riga, who continued coming in the 1990s.[59] It was the relationship between resort and port that proved to be Jurmala's saviour. Nevertheless, the number of daily train connections decreased from the maximum of 120 to about 40 on a summer weekend. Thousands of inhabitants of Jurmala became unemployed, and its population fell from over 60,000 to nearly 55,000.

Jurmala has been recovering slowly. The number of tourists visiting the resort has been steadily rising since 2000. About 100,000 visited in 2005. Finns have led the way, followed by Lithuanians, Germans, Swedes and Estonians. Today, Jurmala is more international and multi-ethnic than it has ever been before. It is also a place of many contradictions, a city where a great inequality of wealth is on display for all to observe. As a result of land speculation, the price of one square metre of land rose in some areas from 10 Latvian latis (about 14 euros) in 1997 to 700 latis (about 1000 euros) in 2006.[60] The real estate dealings have led to the city currently holding the highest property prices in the whole of Latvia. Enormous villas with high fences and surveillance cameras have appeared in the city, while at the same time, some long-term inhabitants of Jurmala, particularly those who are retired, are reduced to virtual poverty and live in crumbling houses. 'I do not like to go to Jurmala any more', says Galina Pagiraine, 'it reminds me of what I cannot afford to buy.' She adds that Jurmala with its reinstituted fences feels very closed up in comparison to the openness of the Soviet era.[61]

Conclusion

The example of the port city of Riga and the resort city of Jurmala illustrates how fairly mundane activities such as the spending of leisure time and being outdoors are socially shaped and controlled by the dominant ideology, be it Russian Imperial, Soviet, Latvian nationalism or post-Soviet market capitalism.

Originally, Jurmala consisted of small fishing villages. The first visitors provided new and welcomed markets for the services and products of the local inhabitants. In the late nineteenth century, the railway connected Jurmala to Riga and other cities in Imperial Russia, while urban inhabitants started to rent houses and increasingly build their own villas by the sea. The seaside was gradually transformed from an all-year-round working environment for poor people into a summertime health resort for mainly upper-class Russians and Germans. It was during the inter-war period that for the first time Latvians became the primary vacationers and developers of Jurmala. Above all, middle-class people from Riga started to spend their summer vacations in the resort.

During the Soviet period, the industrial production of Riga multiplied whilst Jurmala was developed into a major health resort for the socialist tourist industry of the USSR. The existing sanatoriums, hotels and villas were nationalised, and infrastructure was developed to transport and accommodate the Soviet elite and hundreds of thousands of middle-class and better-off working class people. In comparison with its past, Jurmala finally became a relatively socially equal resort because of organised and unorganised mass tourism. After independence, the development of Jurmala became unequal, reflecting the neo-liberal economic policy pursued by the new Latvian governments.

The difference between the two North European coastal towns examined here was clear already in the early part of the nineteenth century. While the city of Riga became a dynamic but contaminated trade and military centre, the city of Jurmala became a resort centre dedicated to purity, beauty and natural healing. With urbanisation and industrialisation in the nineteenth and twentieth centuries, these features became more obvious, thus accentuating the diverging developmental trajectories and contradictory nature of these two cities. Yet, at the same time, the fates of resort and port were connected more strongly than ever before.

Notes and References

1. Richard Kirby and Merja-Liisa Hinkkanen, *The Baltic and the North Seas* (London, 2000). See also Anne E. Gorsuch and Dianne P. Koenker, eds., *Turizm: The Russian and East European tourist under Capitalism and Socialism* (New York, 2006).
2. This study is a result of co-operation between a Finnish research project and Nordic-Baltic oral history network led by Mara Zirnite at the University of Latvia. Field work in Latvia was supported by the Maj and Tor Nessling Foundation for which we are grateful. We also thank Baiba Bela-Krumina and Edmunds Supulis for their valuable help. And Inga Sarma for the kind permission to use the photographs belonging to Jurmala City Museum.
3. Arvis Pope, *Rīgas osta devinos gadsimtos* (Rīga, 2000), pp. 31–40.
4. Pēteris Belte, *Rīgas Jurmalas, Slokas un Kemeru pilsētas ar apkārtni* (1935, published again in 2002 by ULMA), pp. 59–77.
5. Pope, *Rīgas*, p. 15; I. Bernsone, 'Rīgas osta cauri gadsimtiem', in *Daugavas raksti: No Rīgas līdz jurai* , ed. V. Villeruša (Rīga, 1994), pp. 71–89.
6. Andrew Holden, *Environment and tourism* (London and New York, 2002), pp. 26–7.
7. Belte, *Rīgas Jurmalas*, pp. 62, 77, 260.
8. Interview with Inga Sarma, born in Jurmala, 3 August 1959, historian; recorded by Karīna Vasilevska, 21 June 2006.
9. Belte, *Rīgas Jurmalas*, p. 110.
10. Valdis Greiža-Lisovskis, *Jurmala: Vakar, Šodien, rīt* (Avots, 1981) p. 20.
11. Laima Slava, ed., *Jurmala: Nature and Cultural Heritage* (Neputns, 2004), p. 169; Interview with Sarma, 2006.
12. A. Mierina, 'Rīgas teritorija un iedzīvotāji 1860.–1917. gadā', in A. Krastins, ed., *Riga 1860-1917* (Riga, 1978), pp. 7–30.

13. David Kirby, *The Baltic world 1772-1993: Europe's northern periphery in an age of change* (London, 1995), p. 301.
14. Slava, *Jurmala*, p. 237.
15. Interview with Sarma, 2006.
16. Interview with Sarma, 2006; Greiža-Lisovskis, *Jurmala*, p. 81.
17. *Enciklopēdija: Rīga*, ed., P. Jērāns (Rīga, 1988), pp. 73–9.
18. Interview with Sarma, 2006.
19. Interview with Dr Mihaels Malkiels, born 1932, head of 'Jaunkemeri' sanatorium; recorded by by Karina Vasilevska, 9 February 2007.
20. Jurmalas pilsētas darba laužu deputātu padomes izpildu komitejas pirmā sasaukuma protokols, 1. protokols, LVA, 1500, 1, 2, 2.12.1959, p. 121.
21. Greiža-Lisovskis, *Jurmala*, p. 92.
22. Greiža-Lisovskis, *Jurmala*, p. 5.
23. Interview with Sarma, 2006.
24. Slava, *Jurmala*, p. 263.
25. Interview with Gunta Ušpele, born in 1977, head of Jurmala City Tourism Department; recorded by Karīna Vasilevska, 9 June 2006.
26. Interview with Malkiels, 2007.
27. I. Andruce, 'The development of Jurmala resort 1980–1984', Lecture paper, Jurmala, 1984 (Jurmala City Museum Archive), p. 1.
28. Interview with Sarma (2006) and Malkiels (2007).
29. Andruce, 'Development', p. 2.
30. A. Škerbakovs, ed., *Padomju Latvijas arodbiebrību kurorti* (Rīga, 1970), p. 7; see also Anne Gorsuch, '"There is no place like home": Soviet tourism in late Stalinism', *Slavic Review*, 62 (2003), p. 778.
31. A loaf of white bread, for comparison, cost from 0.18–0.22 rouble.
32. Interview with Malkiels, 2007.
33. Slava, *Jurmala*, p 270.
34. Māris Pukītis, 'Jelcina slavas kalējs', in *Kasjauns*, 32 (30 August –5 September 2006), p. 20.
35. Greiža-Lisovskis, *Jurmala*, p. 8.
36. Interview with Ligita Jansone (name changed), born in Jēkabpils, 21 August 1929, chemistry professor; recorded by Karīna Vasilevska, 4 August 2006.
37. Greiža-Lisovskis, *Jurmala*, p. 8. Also local guides routinely present Jurmala as the third biggest holiday resort in the former USSR.
38. http://www.jurmala.lv/lv/home/pilseta/apraksti/statistika/default.aspx (last accessed 4.09.2006); Baumane, *Rīgas Jurmala*, p. 35.
39. Interview with Sarma, 2006.
40. Inta Baumane, 'Rīgas Jurmala, 1940–1959', M.A. thesis, Latvijas Universitāte (2001), p. 83.
41. Interview with Marija Ozola (name changed), born in Riga in 1933, retired teacher and amber artist; recorded by Karīna Vasilevska, 27 July 2006.
42. Interview with Malkiels, 2007.
43. 'Jurmala: kurorts un pilsēta', in *Jurmala*, 28 May 1987, p. 4. According to this article, the resort industry of Jurmala was short of one thousand workers.
44. Interview with Malkiels, 2007; Greiža-Lisovskis, *Jurmala*, p. 8.
45. Interview with Helēna Verdena, born 1938, Aglona, administrative nurse; recorded by Karīna Vasilevska, 23 March 2007.
46. Gorsuch, ' "There is no place like home" ', p. 761.
47. Interview with Galina Pagiraine, born in Rīga, 1951, florist; recorded by Karīna Vasilevska, 5 August, 2006; see Gorsuch, ' " There is no place like home" ', p. 778.

48. Interview with Ozola, 2006, recorded by Karina Vasilevska and e-mail communication on 12 February 2007 with Nataly Kurmangalyeva (name changed), born in 1970, Almaty, Kazakhstan, UN employee.
49. Interview with Pagiraine, 2006.
50. Interview with Sarma, 2006.
51. Dianne Koenker, 'Travel to work, travel to play: on Russian tourism, travel and leisure', *Slavic Review*, 62 (2003), p. 659.
52. Interview with Natālija DemČenko, born in Jurmala, 1962, swimming and gymnastics coach; recorded by Karīna Vasilevska, 9 February 2007.
53. E-mail communication with Nataly Kurmangalyeva, 2006.
54. Interview with Sarma, 2006; interview with Pagiraine, 2006.
55. Interviews with Sarma and Pagiraine, 2006.
56. Interview with Inese ābola, born in Riga 1963, former Jurmala museum employee; recorded by Karīna Vasilevska, June 8, 2006.
57. Interview with Ozola, 2006.
58. Interview with Pagiraine, 2006.
59. Interview with Solveiga Freiberga, Head of the Private Tourism Agency in Jurmala; recorded by Karīna Vasilevska, 9 June, 2006.
60. Dace Plato, 'Visgardākais kumoss Latvijā', in *Diena*, 9 May 2006.
61. Interview with Pagiraine, 2006.

Chapter 12

From Port to Resort: Art, Heritage and Identity in the Regeneration of Margate

JASON WOOD

In October 2006, Blackpool Pleasure Beach put up for sale the cars from its 1935 American-built Turtle Chase. 'Buy your own piece of history on Ebay now', read the advertisement on the Pleasure Beach website.[1] The cars sold for the ridiculously low price of £155 and prompted a hasty trip to Blackpool, armed with a camera and accompanied by my bewildered daughter complaining that 'no-one goes on holiday to see historic rides that don't work'. She had a point. But it started a train of thought. Perhaps people would go on holiday to a place where historic rides *did* work.

Fast forward one year. I am not alone in my thinking. Nick Laister, a leading authority on the British theme park industry, has a plan. His concept, on paper at least, is simple. Acquire a representative sample of classic rides from closed or soon-to-be-closed amusement parks, restore and re-erect them in a single location and create the world's first heritage amusement park dedicated to preserving and operating historic rides. And the place where this dream will come true is, appropriately enough, Dreamland – the former amusement park in the seaside resort of Margate.

Situated on the north coast of Kent, at the extreme end of the Thames estuary, Margate is well known as one of England's earliest seaside resorts and the site of the world's first sea-bathing hospital. From its origins, the resort was particularly popular with middle-class and lower middle-class holidaymakers from London because of its proximity to the capital and relatively good transport links. Rapid expansion in the eighteenth and nineteenth centuries and continued prosperity in the early twentieth century was, however, followed by decline, as happened with so many English resorts, from the 1960s.[2]

Margate seems to have faced greater problems and suffered through a lack of investment more than most resorts in recent decades. Its economic difficulties and social problems were already attracting attention from sociologists and social geographers in the 1980s.[3] Increasingly a byword for faded seaside grandeur, rundown facilities and conflict over its use as a dumping-ground for minorities unwanted elsewhere, the place continues to attract adverse criticism, ridicule and hostile prejudice from sections of the British media and other commentators. The Australian (and honorary Australian) authors of the *Lonely Planet* guide to Britain, originally published in 1995, dismissed the town as freezing and tacky, and concluded with the ultimate and widely reported put-down that God was so depressed looking at Margate that He created Torremolinos.[4] Media attention was again rife in January 2008 when the outspoken singer-songwriter turned political activist Bob Geldof, a long-time resident of Kent, branded the town as unsightly, musing, 'It's a mad mystery, the battle of the ugliness of Margate against the charm of the beaches'.[5] In May 2009, *The Times* carried an aggressive article by Richard Morrison describing Margate as 'a blot on the landscape that could disappear overnight without a murmur of lament' and Dreamland as a 'grotesquely tacky, yet unaccountably "celebrated" amusement park [that] should have been torn down decades ago'. *The Times* published a response from Nick Laister two days later, demolishing Morrison's misplaced rhetoric.[6]

What made Margate unusual among the first resorts was that many of its visitors arrived by sea, initially by single-masted sailing barges or hoys.[7] The harbour and its pier were therefore inextricably linked to Margate's early success as a resort. The harbour itself lay adjacent to the old town, which up until the late eighteenth century retained a modest fishing community. The stone pier or harbour arm, in its latest form built by the engineers John Rennie and William Jessup, dates from 1815 (see Figure 12.1). This year also saw the introduction of the first regular steamer service from London, providing a quicker and more comfortable method of transport. As the pier could be used only at high tide, a wooden jetty known as Jarvis' Landing Place was erected in 1824 immediately to the east of the harbour. This enabled additional steamers to disembark and embark, and by 1835, passenger numbers had risen five-fold to 109,000 (see Figure 12.2).[8] The jetty was later rebuilt in iron to designs by the celebrated seaside pier engineer Eugenius Birch. It opened in 1855 and was in fact the first of Birch's many commissions. In 1877, the jetty was extended with the addition of a hexagonal pier head, but virtually all of the structure was lost to a storm in 1978.

The arrival of the railway and construction of two stations – Margate Sands in 1846 and Margate West in 1863 – shifted the development of holiday accommodation and entertainment further west, away from the old town and harbour. The largest and most important of these late nineteenth-century developments was the site that was to become Dreamland.[9]

The decline of the harbour and the buildings supporting the port and fishing activities began much earlier than the decline of the resort infrastructure. The area became increasingly derelict and much of it,

Figure 12.1 The stone pier or harbour arm, a Grade II listed building. Margate's early prosperity depended on the harbour as the base for a fishing fleet and as an outlet for agricultural produce.

including the imposing Hotel Metropole, was cleared away in the late 1930s as part of the so-called Fort Road Improvement scheme, to be replaced by a short section of dual carriageway. By comparison, the decline of Dreamland was relatively late, with the amusement park site being largely cleared in 2002 (with a dwindling number of rides continuing until 2006) and the adjacent Dreamland cinema complex only closing in 2007.[10]

The historical importance to the development of Margate of these two key areas, the harbour and the Dreamland site, is now being recognised, and their decline is beginning to be reversed through a major programme of investment in art and heritage that seeks to marry regeneration, culture and identity to deliver exciting new visitor attractions and associated redevelopment.

Art-Led Regeneration – Turner Contemporary

The harbour area is currently being transformed by the construction of a gallery for Turner Contemporary, a visual arts organisation that celebrates the association between Margate and one of its most famous past residents, the artist J.M.W. Turner. It does this through a varied

Figure 12.2 Passengers awaiting the steamer service on Jarvis' Landing Place, 1848.

programme of exhibitions and events that 'promotes an understanding and enjoyment of historical and contemporary art'.[11] Turner (1775–1851) lived in Margate as a child for a short period and returned there in his early fifties for summer weekends, attracted by the quality of its light, its seascapes and the charms of his landlady Mrs Sophia Booth. More than 100 of his works, including 30 large canvases, were inspired by the Kentish coast.[12]

The idea of an art gallery or 'Turner Centre' was mooted as early as 1994 but has been fraught with difficulties and controversy ever since. The prestigious design competition was originally won by the Norwegian architects Snøhetta in partnership with the London-based architect Stephen Spence. Their proposal envisaged the gallery daringly perched on part of the harbour arm, but this ultimately unrealistic scheme had to be abandoned in February 2006 because of technical problems and escalating costs of at least £25 million.[13] The more conventional land-based gallery presently being built, designed by the RIBA Stirling Prize–winning architect David Chipperfield, will open in 2011 for a total cost of £17.4 million, with funding from the Kent County Council, the Arts Council England, the South East England Development Agency and private donations.[14]

The brief for the gallery called for an inspiring architectural design that took maximum advantage of the outstanding seafront location, particularly the uninterrupted views out to sea and spectacular sunsets across the harbour and bay. Such a prominent and exposed position meant that the building would have to endure challenging physical conditions caused by high winds and high seas. The building would also have to be sensitive to the setting of two listed buildings – the harbour arm and Droit House, originally built as a customs building in 1812 (see Figure 12.3). The harbour arm and Droit House were both prominent

Figure 12.3 Droit House, a Grade II listed building, originally built by the Margate Pier and Harbour Company.

features of the urban landscape in Turner's time, and the gallery itself occupies the same location as the lodging house in which Turner stayed whilst in Margate. The size and shape of the new building was another important consideration since the brief required a design that would help restore and strengthen the historical and physical relationship between the old town and the waterfront. There were site restrictions too, in that the continued operation of the adjacent RNLI lifeboat station had to be maintained.[15]

In its final form, the building will comprise a cluster of six identical, interlocking, rectangular blocks, positioned on a plinth and laid out over two floors. The suite of first-floor galleries will be lit by natural 'maritime' light from north-facing rooflights and smaller skylights in the repeating monopitch roofs. The façade of inch-thick sheets of glass will give the building texture and a milky white appearance (see Figure 12.4). There will be some panoramic views out to sea as well as the opportunity to view the town and the bay.[16] In Chipperfield's own words, 'our building will look out to the sea, connect itself to the town and capture the same unique light that inspired Turner'.[17]

Related development has already included the conversion of Droit House to a temporary visitor centre and exhibition venue, as well as the

Figure 12.4 The design for Turner Contemporary celebrates Turner's fascination with light. 'From the spacious naturally lit galleries to its opaque glass exterior, the building will absorb and reflect light to create a distinctive and inspirational building.'

removal of the dual carriageway to help reintegrate the old town and the harbour area. The harbour arm re-opened to the public in May 2008 after refurbishment to create an art gallery and artists' studios.[18] The Turner Contemporary Project Space, an outreach programme in a former Marks & Spencer building in the old town, ran until September 2009. Future developments in the harbour area include a mixed-use scheme adjacent to the gallery on the so-called Rendezvous site. This originally proposed a 120-bed four-star hotel (Margate is in desperate need of premium hotel accommodation), 150 apartments and some commercial development.[19] However, many considered this scheme to be of mediocre quality and, following the withdrawal of the developer because of the economic downturn, work has now started on a revised masterplan with a wider remit to include regeneration sites stretching up Fort Hill to the Winter Gardens, Lido and beyond.

Since as early as 2001, Turner Contemporary has been successfully developing an audience for visual arts in Margate and East Kent and, importantly, demonstrating an active commitment to working with the local community. The vision has always been one of learning and participation, reaching out to the community to reinvigorate a sense of local pride and identity and to enrich people's lives. In delivering its varied creative programme, Turner Contemporary has worked particularly with teachers and artists to enhance the art curriculum for schoolchildren, while also engaging with adult learners and diverse groups of people including migrant communities. A particular target has been those people living in the most socially and economically deprived areas, who tend to feel excluded from or indifferent to art and art galleries.[20] For example, the Cultural Ambassadors project has provided opportunities for people to find out more about the arts and benefit from and participate in the regeneration of their area. A five-year collaboration with the University for the Creative Arts at Canterbury has helped provide educational opportunities for young students, many of whom have been the first generation in their families to experience further and higher education. The syllabus offers a model of good practice, spawning other partnerships and collaborations.[21]

The exhibition programme for the new gallery is now taking shape with the first 18 months already planned. As the gallery will not have a permanent collection, this will comprise a series of three or four temporary exhibitions per year revolving around three core interlocking strands: loaned works by Turner, historic art (post-1750) to provide a context for the Turners, and contemporary work. There will always be a Turner presence, but the emphasis will be on a strong contemporary programme.[22]

It is anticipated that the exhibitions will put Margate on the 'arts map' and act as a catalyst for reinvention of the old town as a distinctive arts and creative quarter, as well as adding significantly to the cultural provision in Kent. The aim is to make Margate a year-round arts destination and help attract a new, different, more up-market visitor base nationally and internationally.[23] In this respect, Margate is following the lead of the Cornish towns of St Ives and Newlyn, which have long traded on their artistic links and, more recently, Folkestone with its triennial exhibition of contemporary artworks.[24] The prospect of a major new gallery is already attracting new investments, businesses and residents to the area and making a significant difference to the regeneration of Margate and East Kent in general. In particular, cafés, restaurants, shops, galleries, art workspaces, residential studios and commercial facilities, such as the Margate Media Centre, have begun to populate the old town and its large concentration of high-quality historic buildings.

Key to these developments has been the establishment of the Margate Renewal Partnership, comprising relevant agencies and stakeholders, to lead and coordinate the wider regeneration and transformation of Margate. The Partnership has already successfully spearheaded a number of schemes and invested heavily in the public realm and cultural activities as part of its campaign to change perceptions and to raise the profile of the town.[25] Its current focus is on securing a future for the Dreamland site where the emphasis is on creating a leisure attraction that celebrates Margate's seaside heritage and popular culture.

Heritage-Led Regeneration – Dreamland Heritage Amusement Park

The Dreamland site has been a leisure venue for over 140 years. In 1867, the railway caterers Spiers and Pond opened the Hall-by-the-Sea entertainment complex in an obsolete railway station building adjacent to the Margate Sands terminus. The Hall was originally used for concerts and dances but was soon sold to the developer Thomas Dalby Reeve, then mayor of Margate, who also acquired land to the rear. In 1874, ownership of the site passed to the self-ennobled circus entrepreneur 'Lord' George Sanger,[26] who refurbished the building for dual use as a ballroom and restaurant and developed the land to include ornamental pleasure gardens, a small lake, a 'ruined abbey' folly, a menagerie and later a roller-skating rink. The old railway building was eventually replaced in 1898 with a *fin-de-siècle* mirrored ballroom designed by Richard Dalby Reeve. In 1919–1920 the complex was again extended by the new owner, the entertainment entrepreneur John Henry Iles.[27] It was Iles who opened the American-style amusement park and renamed the site Dreamland (the name probably being taken from a short-lived but enduringly famous Coney Island attraction). In 1923, the existing ballroom was put to use as a 900-seat cinema, and a new ballroom was created on the site of the roller-skating rink to the rear. A large new restaurant was also constructed. In 1935, these structures were substantially replaced by a large, purpose-built 2200-seat cinema and multi-use entertainment complex providing a new and enticing entrance to the park. This was designed by Julian Rudolph Leathart and W. F. Granger, important cinema architects of the period, with some of the interior decorations and furnishings by John Bird Iles (the son of John Henry). The building's seafront elevation was furnished with a distinctive fin tower carrying vertical lettering – the first time such a feature had been used in cinema design in the United Kingdom – introducing what would go on to become a prominent characteristic of many Odeon façades. As well as the cinema, the complex also included the existing ballroom, together with new bars, cafés and restaurants.[28]

The 20-acre amusement park opened in 1920 with a giant Scenic Railway roller coaster as its centrepiece. The park drew 1.5 million visitors during its initial year of operation. Other rides and attractions in the first full season included a Joy Wheel, Cake Walk, Helter Skelter, Miniature Railway, Whip, Tumble Bug, Lunar Ball, Haunted Castle, Hall of Mirrors and House of Nonsense. In successive years, more rides were brought into the park, including a Caterpillar, River Caves, Racing Coaster, Rapids, Over The Falls, Motor Boats, Brooklands Racers (see Figure 12.5), Galloping Horses roundabout, Whirlwind, Octopus, Sky

Figure 12.5 A Dreamland publicity shot of the Brooklands Racers in the late 1940s.

Wheels, Rock and Roll House, Go Karts, Satellite, Jets and Whirl-a-Boats. During the 1970s, a number of the long-established rides were dismantled and replaced by more modern rides including a Paratrooper, 20,000 Leagues under the Sea, Astroglide, Orbiter, Tip-Top, Swirl, Cyclone, Water Chute, Dodgems and Big Wheel. Changes of owners in the 1980s and 1990s saw the arrival of new thrill rides including a Looping Star, Pirate Ship, Traum Boot ride called Mary Rose, Chair-o-Plane, Meteor gravity ride, Enterprise, Log Flume and Wild Mouse.[29]

Over the years, Sanger's original ornamental pleasure gardens were transformed with decorative lighting and illuminated figures into the Magic Garden, later converted to the Safari Zoo. Clustered around the park were also numerous restaurants, cafés, stalls and sideshows, some contained in relocated and renovated First World War aircraft hangars, later replaced by purpose-built arcades, including one in the shape of the liner Queen Mary (see Figure 12.6). Important among these additional buildings were a coach station and buffet/cafeteria designed in the modernist style by C.F.S. Palmer in 1928–1930.

Today, the only amusement park ride to survive is the Scenic Railway constructed in 1919–1920 (see Figure 12.7). It is the oldest surviving roller coaster in the United Kingdom and the third oldest of its type in the world.[30] It was also the first amusement park ride in the United Kingdom to be listed (Grade II in 2002).[31] The magnificent Dreamland cinema

Figure 12.6 An aerial view of Dreamland in July 1952.

Figure 12.7 The Dreamland amusement park site currently remains vacant except for the Scenic Railway roller coaster.

complex also survives (see Figure 12.8). Although the building has undergone alterations and changes of use (including twinning of the cinema circle and conversion of the stalls to a theatre and later bingo hall) much of the original, high-quality art deco decoration and fittings are preserved, and its enhanced significance has recently been recognised by the upgrading of its listing to Grade II*.[32] The rear of the cinema complex still preserves some elements of the 1923 ballroom (converted to squash courts in the 1970s and now in very poor condition), including a section of the original rustic enclosure wall of Sanger's menagerie. Further remains of the menagerie enclosure and three bear display cages were uncovered in 2008 and are now also listed.[33] Locating the heritage amusement park at Dreamland will therefore secure a sustainable future and improve the setting of this important group of listed buildings.

The impetus behind the scheme is the unprecedented rate of closures of amusement parks in the United Kingdom and, consequently, the increasing number of historic rides under threat. Fifteen traditional seaside parks and inland theme parks have shut down since 1999, with more closures expected in the coming years. At those parks that remain, major changes are being planned which will further erode what heritage survives. These losses and changes are not without controversy. In Southport, for example, popular concern for the Pleasureland

Figure 12.8 The Dreamland cinema complex was the highest structure on Margate seafront for nearly 30 years. Since 1962, it has been dwarfed by the Arlington House tower block.

amusement park led to public protests and failed attempts at heritage designation to prevent the disposal of a rare Cyclone roller coaster built in 1937.[34]

Despite a long tradition of open-air museums and heritage parks in the United Kingdom and elsewhere, there is no such place dedicated to amusement park rides anywhere in the world. Blackpool Pleasure Beach and Kennywood Park, Pittsburgh (United States), although still retaining a large number of historic rides, are not equivalents, while the Dingles

Fairground Heritage Centre at Lifton in Devon is very different in that it focuses exclusively on travelling rides and is located on a farm. The Hollycombe Steam Fairground at Liphook in Hampshire and the living industrial archaeology museums at Blists Hill, Beamish and in the Black Country are more closely analogous to the heritage amusement park proposal and lend the Dreamland project the respectability it may need through precedent.[35] A key point here is that the historic rides will be restored not as museum pieces but as active rides to be experienced for real. The concept is to offer physical, intellectual and emotional access with an appeal to tradition, authenticity, nostalgia and identity.

A further justification is that amusement parks still draw a wide audience. Dreamland, in its heyday, attracted 2 million visitors annually, and Blackpool Pleasure Beach, until recently, regularly topped the list of the United Kingdom's free attractions, with 5.5 million visitors in 2007.[36] The closures of amusement parks in the last decade have not been due to lack of demand; the opposite has been the case. During the property boom of the late 1990s, land increased in value, and the owners of amusement parks were encouraged to sell for housing and retail development. The irony is that many of these sites are now derelict, because the firms that bought them have gone out of business. In other words, the Dreamland heritage amusement park is not a case of trying to preserve something that people have lost interest in. In fact, it is anticipated that the park will generate 700,000 annual visits in its first year of operation, matching the estimate for 2001,[37] with the main target audience being family orientated, particularly the 'baby-boomer' generation with a longing for the amusement parks of their youth.

The arguments regarding the impetus, precedent, concept and demand have been made and accepted in principle. Redeveloping the Dreamland site, however, and the adjacent and equally neglected 1960s shopping centre and car park around the Arlington House tower block represents a considerable challenge. A planning brief for Dreamland was approved by Thanet District Council in February 2008. This proposed a mixed scheme, with just over half of the site, including the major heritage assets, retained as part of an amusement-based destination. The rest of the development, with improved links with the beach-front, will include commercial and residential buildings.[38] There is no doubt that the public is behind the heritage amusement park initiative. Consultation on the planning brief clearly demonstrated support for retaining the Scenic Railway (over 92%), more than half of the site as an amusement park (over 87%) and the cinema complex for leisure uses (over 85%).[39] A more recent online survey by Visit Kent in April 2009 had replies from over 3300 people, with 93% in favour of the Dreamland project. What is more, an influential report by English Heritage's Urban Panel unequivocally endorsed the vision for Dreamland, arguing forcefully that it be promoted 'with urgency and drive and without unnecessary burdens'.[40]

Proposals to create the world's first heritage amusement park were initially put forward in April 2007 by the Dreamland Trust, a not-for-profit organisation that grew out of the vigorous and widely supported Save Dreamland Campaign.[41] Nick Laister, Chair of the Trust, founder of the Campaign and instrumental in securing the listing of the Scenic Railway, first set out the case in a preliminary report on the availability of historic rides and an assessment of space requirements.[42] Based on this, a masterplan for the site adjacent to the Scenic Railway was drawn up; the latest version, launched in March 2009, being by architects Levitt Bernstein and the internationally renowned amusement park designer Jean-Marc Toussaint. This masterplan details the size, shape and indicative layout of the proposed 9-acre park and the planned uses for the cinema complex. In the meantime, the Trust has already started to acquire, systematically dismantle, transport and store several historic rides from Pleasureland in Southport – Caterpillar, Chairlift, Ghost Train, Flying Scooters, Wild Mouse, Meteorite, River Caves (boats and mechanical parts only), Mirror Maze, Haunted Swing and Fun House machines[43]; from Ocean Beach in Rhyl – Water Chute (trains and mechanical parts only); and from Blackpool Pleasure Beach – Whip and Junior Whip. Some of these 13 rides are the last surviving examples of their type in the United Kingdom. A further 15 rides from various locations in the United Kingdom and the United States have been identified for potential acquisition, and the current owners have been approached. Of these, the highest priorities are a Corbière Ferris Wheel, Gallopers, Miniature Railway and Helter Skelter.

It is envisaged that the cinema complex, once restored and refurbished, will again form the main entrance to the park, as well as providing imaginative interpretation space to showcase the rich history of the Dreamland site through a variety of objects, relics, memorabilia, films and archives. The intention is that this exhibition will establish the foundations of a national centre to explore the history, heritage and popular culture of amusement parks, their association with seaside resorts and the entertainment industry and their relevance in British society and beyond. In a complementary initiative, there are also plans to celebrate the British seaside through a programme of festivals on the theme of post-war youth culture embracing fashion, film, photography, graphic design and music. This links strongly to Margate's identity and continuing appeal as a place for regular gatherings and rallies. In due course, the cinema complex could provide a home for a national centre for youth street culture, interpreting and celebrating this unique aspect of British life from the Teddy Boys of the 1950s to the Chavs of the early twenty-first century.[44] As for the large auditorium, a commercial partner will be sought to transform the space into a flexible venue for a range of performing arts and other activities such as conferences. This, of course, would be entirely appropriate as the original building, and its earlier incarnations, functioned as multi-purpose entertainment complexes.

Work has already begun on relevant archives, oral history and audience development. A wealth of material held at the National Fairground Archive at the University of Sheffield, Blackpool Pleasure Beach and elsewhere is beginning to be catalogued and researched. Personal reminiscences of key individuals with close associations with the Dreamland site are being recorded, while peoples' memories and emotional attachment to the amusement park are being captured at innovative and celebratory public history events such as *Dreamcoaster* (see Figure 12.9) and *I Dream of Dreamland*. Repair and restoration of the various historic rides will be partly based on information derived from the archives and oral testimony. An idea has also been floated to construct a replica of a lost historic ride from Dreamland and to use this as an outreach project to build the capacity of the local community to participate actively and connect more fully with their heritage. In addition, the amusement park will provide an attractive medium for

Figure 12.9 The Dreamcoaster exhibition invited members of the public to donate materials, lend memorabilia and participate in building a scaled-down replica of the Scenic Railway. People were encouraged to share memories of Dreamland and the Scenic Railway and pin them to the replica.

extending volunteer involvement and strengthening educational links with local schools and colleges.

A potential setback, however, was an arson attack on the Scenic Railway in April 2008. This received national media coverage with concomitant exaggerated reporting and accusations. Fears that the roller coaster was permanently lost were wide off the mark and could be easily quashed once the true extent of the damage was established. About 20%–25% of the ride's footprint, including the main lift hill, was destroyed (see Figure 12.10). Everything else was deemed structurally sound and restoration to full working order was considered relatively straightforward.[45] Regrettably, the fire started in the workshop which contained the trains. Thankfully, however, some disused trains of exactly the same date and gauge have been found in Budapest and negotiations to bring them to Margate are now in hand.

The episode, although unwelcome, nevertheless had the effect of galvanising the various interested parties into constructive dialogue so that the project was able to move forward as a joint venture. A *Memorandum of Understanding* was agreed in 2009 between the landowner and developer, Margate Town Centre Regeneration Company Ltd.,[46] and the key stakeholders, Thanet District Council, the Dreamland Trust and the Margate Renewal Partnership. Together, these four organisations comprise the Dreamland Client Group. The *Memorandum* foresees 51% of the entire site being transferred to the Dreamland Trust and the guarantee of £4 million of developer funding through a *Section 106* agreement as part of the planning permission for the redevelopment of the

Figure 12.10 The Scenic Railway fire destroyed much of the central part of the ride, including some of the station.

remaining 49%. A further £3.7 million has been secured from the Sea Change programme, with significant contributions also from the Heritage Lottery Fund and others.[47] Assuming the necessary funding is in place, phase one of the project, including most of the heritage amusement park, shell repair of the cinema complex and reopening of the ground floor entrance to the park, is scheduled to be completed by Easter 2012, with phase two, including full restoration of the cinema complex and completion of the park, following three years later.

Conclusion

Margate is the embodiment in media circles of a down-at-heel resort – a favourite symbol of seaside economic collapse. There is no doubt that it is a severely deprived area in need of serious regeneration and a new image (see Figure 12.11). The potential has clearly been recognised, and investment is now starting to pay dividends, but there remains much to do. Margate Renewal Partnership's vision statement for the town is resolutely upbeat, promising a more confident, prosperous and exciting future:

> By 2015, Margate will become a dynamic, thriving and successful town. It will be a major hub and driving force of creativity and culture that excites and inspires residents and visitors alike. It will

Figure 12.11 Margate – in search of a new direction.

embrace and celebrate its traditions as a place of relaxation, leisure and seaside fun.[48]

It is an ambitious aspiration and target. In August 2009, with the British economy still in the throes of recession, figures were released by the Local Data Company showing that Margate statistically had the worst rate of shop closures in England and Wales, with 25% of the town's retail space unoccupied.[49] Undaunted, Thanet District Council began offering the vacant premises to local artists to display their work in an attempt to improve the appearance of the high street. Meanwhile, supporters of the Margate Renewal Partnership's re-branding initiatives refuse, not unreasonably, to be deflected from seeking to build on the town's arts credentials, social history and popular culture and revitalising areas to encourage modern retail and a new lifestyle.[50] English Heritage also recognises the opportunity in Margate for 'a single integrated programme in which art, creativity and heritage combine in a new sustainable business model', a model of symbiosis that could be exported to similarly challenging seaside resorts.[51]

Although part of a much broader creative strategy for the town, Turner Contemporary and the Dreamland heritage amusement park are central to this vision and emblematic of the change and progress that is being achieved. Both are anchors for Margate's regeneration, aimed at reversing the acute decline in the town's fortunes over the last few decades. Collectively, they have the capacity to play important catalytic and complementary roles in enhancing the status of Margate as an arts and heritage destination. They are bound together in a cultural narrative that residents, visitors and investors can comprehend. There is also scope for cooperation between the two projects.[52] Both will provide significant additional tourism opportunities, help bolster civic pride and make a positive contribution to the growth of the area's economy. Importantly, they value the character and traditions of Margate and its collective memory, heritage and identity as both a resort and a port.

Acknowledgements

I am grateful to Victoria Pomery of Turner Contemporary, Nick Laister of the Dreamland Trust and Derek Harding of the Margate Renewal Partnership for information and comments.

Notes and References

1.   http://www.blackpoolpleasurebeach.com/index.php accessed 27 Oct. 2006; http://cgi.ebay.co.uk/Pleasure-Beach-Turtle-Chase-Amusement-Park-Ride-Cars_W0... accessed 27 Oct. 2006. The Turtle Chase was built by the Traver Engineering Company at Beaver Falls, PA, and was originally known as the Tumble Bug. It was reconstructed in 1958 taking on the Turtle Chase appearance and name.

2. For the most recent study of the town's history and heritage, see Nigel Barker, Allan Brodie, Nick Dermott, Lucy Jessop and Gary Winter, *Margate's Seaside Heritage* (Swindon, 2007). See also the chapter by Allan Brodie and Gary Winter in this volume.

3. P. Cooke (ed.), *Localities* (London, 1989), Chapter 5; C. Pickvance, 'Council economic intervention and political conflict in a declining resort: Isle of Thanet', in M. Harlow, C. Pickvance and J. Urry (eds.), *Place, Policy and Politics: Do Localities Matter?* (London, 1990), pp. 132–49.

4. Richard Everist, Bryn Thomas and Tony Wheeler, *Britain: a Lonely Planet travel survival kit* (Hawthorn, Victoria, 1995), p. 284.

5. See, for example, Amol Rajan, 'Saint Bob's latest rant creates a storm in 'ugly' town of Margate', *The Independent*, 19 Jan. 2008: http://www.independent.co.uk/news/uk/this-britain/saint-bobs-latest-rant-creates-a-storm-in-ugly-town-of-margate-771220.html accessed 20 Jan. 2008.

6. Richard Morrison, 'What's wrong with Margate', *The Times*, 13 May 2009: http://www.timesonline.co.uk/tol/comment/columnists/richard_morrison/article6274667.ece accessed 3 Sept. 2009. For a much more upbeat reflection on the Dreamland site, see Iain Aitch, 'Rollercoaster ride into the past', *Sunday Express*, 29 Mar. 2009: http://www.express.co.uk/posts/view/91844/Rollercoaster-ride-into-the-past- accessed 3 Sept. 2009.

7. J. Whyman, 'Water communications to Margate and Gravesend as coastal resorts before 1840', *Southern History*, 3, 1981, pp. 111–38.

8. Whyman, 'Water communications'.

9. F. Stafford and N. Yates (eds.), *The Later Kentish Seaside, 1840–1974* (Stroud, 1985).

10. Nick Evans, *Dreamland remembered: 90th anniversary edition – celebrating Margate's famous amusement park* (Whitstable, 2009).

11. http://www.turnercontemporary.org/ accessed 19 Aug. 2009.

12. For an accessible account of Turner's life, see Peter Ackroyd, *J.M.W. Turner* (London, 2005), especially p.114ff for Turner's visits to Margate from 1827. More recently Margate has been recognised for its links to Tracey Emin, shortlisted for the Turner Prize in 1999, whose early works were inspired by her experiences growing up in the town.

13. For background information, see Kent County Council, *A Turner Centre for Kent* (1999); Tim Mason, *Dreaming with open eyes: a report on the cultural content and focus of a proposed Turner Centre for Margate* (report produced for Kent County Council, 2001); Victoria Pomery, 'Selling culture to Margate', *Context*, 80, July 2003, pp. 15–16. A legal dispute between Snøhetta and Kent County Council over 'wasted' design costs from the failed scheme was settled in favour of the Council in Oct. 2009: http://www.architectsjournal.co.uk/news/daily-news/6-million-out-of-court-settlement-for-snhettas-failed-margate-centre/5209727.article accessed 17 Nov. 2009.

14. David Chipperfield Architects Ltd. was appointed in July 2006 and building work began on site in December 2008. For a discussion of the project since 2006, see Jason Wood, (2010) 'Contemporary history in the making', http://www.turnercontemporary.org/about/?p = 203 accessed 18 June 2010.

15. *Turner Contemporary: general design brief for new gallery* (2006).

16. *Turner Contemporary: Stage D report* (Nov. 2007).

17. David Chipperfield spoke about his revised designs at a public meeting at the Theatre Royal, Margate, 16 Oct. 2007. See also http://www.turnercontemporary.org/design/?p = 122 and http://www.turnercontemporary.org/

uploaded_documents/David%20Chipperfield%20Public%20Presentation%20 at%20The%20Theatre%20Royal%20.ppt accessed 19 Aug. 2009.

18. http://www.margateharbourarm.co.uk/ accessed 20 Aug. 2009.

19. http://www.thisismargate.co.uk/pdf/Rendezvous_site_press_release_7 march08.pdfaccessed 3 Sept. 2009. The Rendezvous site was originally occupied by swimming baths and the Fort Arcade.

20. For more information, see http://www.turnercontemporary.org/learn/ accessed 3 Sept. 2009.

21. Gillian Wilson, *Life and art: work related learning in Margate* (University for the Creative Arts 2009).

22. *Turner Contemporary programme strategy* (2010).

23. See, for example, Locum Consulting, *Margate destination strategy* (May 2006), commissioned by Thanet District Council to examine latent development opportunities and a more sustainable economy.

24. On this theme, see Rachel Cooke, 'Can art put new heart into our seaside towns?', *The Observer*, 16 Aug. 2009: http://www.guardian.co.uk/artanddesign/ 2009/aug/16/art-in-british-seaside-towns accessed 3 Sept. 2009.

25. http://www.margaterenewal.co.uk/default.aspx accessed 3 Sept. 2009. The Margate Renewal Partnership includes, among others, representatives from Kent County Council, Thanet District Council, the South East England Development Agency, English Partnerships, the Government Office for the South East, Arts Council England, the Heritage Lottery Fund and English Heritage.

26. For Sanger, see his autobiography *Seventy years a showman* (London and Toronto, 1926).

27. For Iles, see Oxford DNB: http://www.oxforddnb.com/index/101048777/ Henry-Iles, accessed 18 June 2010.

28. John Hutchinson, *A dream came true* (Margate, 1995). See also Richard Gray, *Cinemas in Britain: one hundred years of cinema architecture* (London, 1996), p. 93. Dreamland was Leathart and Granger's last cinema scheme as the partnership was dissolved in 1937.

29. Evans, *Dreamland remembered*. Several rides and sideshows appeared in the groundbreaking short documentary *O Dreamland* by British film director Lindsay Anderson (made in 1953, released 1956): http://www.screenonline.org.uk/film/ id/438978/ accessed 3 Sept. 2009. The Scenic Railway and some of the later thrill rides featured in *The Jolly Boys' Outing* – a Christmas Special of the classic sitcom *Only Fools and Horses* (BBC, 1989). The park also featured prominently in Pawel Pawlikowski's film *Last Resort* (2000), about a young Russian immigrant seeking asylum in England.

30. The two earlier surviving scenic railways are at Luna Park, Melbourne (1912) and Tivoli Gardens, Copenhagen (1914). Hundreds were built in the USA and about 40 in the UK. There are no examples, of any period, surviving in the USA. The only other scenic railway still operating in the UK is at the Pleasure Beach, Great Yarmouth (1932, although much altered).

31. For the listed building description, see http://www.heritagegateway.org.uk/ Gateway/Results_Single.aspx?uid = 488465&resourceID = 5   accessed   3 Sept. 2009. An application to up-grade to II* is in process. See also Nick Laister and David Page, *Request for spot listing of the Scenic Railway roller coaster, Dreamland Amusement Park, Margate, Kent: submission to the Secretary of State for Culture, Media and Sport* (May 2001). The only other listed amusement park ride in the UK is the rare 1929 Water Chute at Hull's East Park (Grade II).

32. For the listed building description, see http://www.heritagegateway.org.uk/ Gateway/Results_Single.aspx?uid = 441128&resourceID = 5accessed 3 Sept. 2009. The cinema complex is the most important building in the Margate Seafront Conservation Area.

33. Listing was confirmed in Feb. 2009: see http://www.savedreamland.co.uk/ accessed 3 Sept. 2009. Palmer's buffet/cafeteria building also survives, but in a much altered form and unlisted.

34. 'Southport roller coaster (Cyclone) destroyed', 14 Sept. 2006: http:// www.southportforums.com/forums/showthread.php?threadid = 50365006 accessed 10 Sept. 2009; 'Cyclone: second demonstration takes place', 20 Sept. 2006: http://www.southportforums.com/forums/printthread.php?t = 50365239&pp = 40 accessed 10 Sept. 2009.

35. For Blackpool, see John K. Walton, *Riding on rainbows: Blackpool Pleasure Beach and its place in British popular culture* (St Albans, 2007); and for Kennywood, David P. Hahner Jr., *Images of America: Kennywood* (Charleston SC, 2004). For a discussion on the authenticity and identity of the historic amusement parks at Blackpool and Kennywood, see Scott A. Lukas, *Theme Park* (London, 2008), pp. 130–33; for Dingles, see http://www.fairground-heritage.org.uk; for Hollycombe, see http://www.hollycombe.co.uk/.

36. http://www.enjoyengland.com/Images/Visits%20to%20Visitor%20At tractions%20Survey%2007%20-%20Top%20Attractions%20-%20Top%20 Free%20Attractions_tcm21-170660.pdf accessed 28 Sept. 2009.

37. http://www.enjoyengland.com/Images/Major%20UK%20Visitor%20Attrac- tions%202002_tcm21-170879.pdf accessed 28 Sept. 2009.

38. http://www.thisismargate.co.uk/pdf/7Dreamland%20Margate%20SC_%20 Dreamland%20Planning%20Brief.pdf accessed 28 Sept. 2009. A separate planning brief for the Arlington site was adopted in Oct. 2008 following overwhelming support for exterior improvements to the tower block and plans for new shops and cafes facing the seafront: http://www.thanet.gov.uk/ council__democracy/consultation/arlington_brief_adopted.aspx accessed 28 Sept. 2009.

39. http://www.thanet.gov.uk/council__democracy/consultation/dreamland_ plan.aspx accessed 28 Sept. 2009.

40. *English Heritage Urban Panel: Margate review paper* (20 Apr., 2009).

41. For background information on the crucial role played by the Save Dreamland Campaign in raising the profile of the site and their continuing success in shaping plans, see the regularly updated news pages at http:// www.savedreamland.co.uk/ and http://www.joylandbooks.com/sceni- crailway/heritageamusementpark.htm. The project's URL is www.dream landmargate.com.

42. Nick Laister, *Proposed heritage amusement park, Dreamland, Margate: ride availability, concept plan and business plan* (May 2007).

43. For information on these rides, see Stephen Copnall, *Pleasureland memories: a history of Southport's amusement park* (St Albans, 2005).

44. An idea first put forward by Nick Dermott and Keith Hayward, *Dreamland Cinema, Margate: Centre for British Youth Cults* (Sept. 2007).

45. Condition reports by Jacobs Engineering Group Inc. and English Heritage. The ride was successfully rebuilt following fire damage on two previous occasions in 1949 and 1957. See also 'Calls to rebuild the Scenic Railway', Thanet District Council news release (8 Apr., 2008).

46. *Dreamland: a development by Margate Town Centre Regeneration Company* (2008) is the latest in a series of developer/architect-led prospectuses.

47. Sea Change is a capital grants programme designed to invest in the most deprived seaside resorts in England and contribute to sustainable social and economic regeneration. The programme was instrumental in funding an extensive feasibility study into proposals for the Dreamland site: http://www.cabe.org.uk/sea-change/margate accessed 29 Sept. 2009 and http://www.thisismargate.co.uk/PDF/Margate%20Renewal%20Board%205_3_09.pdf accessed 29 Sept. 2009. For details of the main grant application and supporting documents including the business plan (by Locum Consulting), outline design (by Levitt Bernstein), conservation statement (by the Prince's Regeneration Trust), etc, see http://www.thisismargate.co.uk/read_all_about_it/i_dream_of_dreamland.aspx accessed 29 Sept. 2009. For the award announcement, see http://www.cabe.org.uk/news/sea-change-for-seven-resorts accessed 16 Nov. 2009.
48. http://www.margaterenewal.co.uk/our_vision.aspx accessed 3 Sept. 2009.
49. Michael Savage, 'How the recession turned Margate into a ghost town', *The Independent*, 15 Aug. 2009: http://www.independent.co.uk/news/uk/home-news/how-the-recession-turned-margate-into-a-ghost-town-1772517.html accessed 4 Sept. 2009.
50. See, for example, *A cultural vision for Margate: Creative Margate ten year delivery plan* (2008) and its six visioning themes. Some omens, however, remain daunting – Margate Museum closed at the end of March 2009 and is still mothballed: http://home.btconnect.com/margatemuseum/ accessed 17 Nov. 2009.
51. Andy Brown, 'A creative future for seaside resorts: Margate, Turner and beyond', *Conservation Bulletin*, 62, Autumn 2009, pp. 37–9.
52. For example, as part of its *Time of Our Lives* project, Turner Contemporary is planning a major exhibition entitled *Teenage*, to be held in the new gallery in its opening year. This artist-led intergenerational project will enable local people to explore themes arising from their experiences of being a teenager, both now and in the past. There are clear synergies with the archives and oral history work being undertaken by the Dreamland Trust and the proposed youth street culture programme. See: http://www.turnercontemporary.org/uploaded_documents/coordinator%20job%20description.doc accessed 28 Sept. 2009.